Contingent Valuation of Yangtze Finless Porpoises in Poyang Lake, China

Yanyan Dong

Contingent Valuation of Yangtze Finless Porpoises in Poyang Lake, China

Springer

Yanyan Dong
Economics
Helmholtz Center for Environmental Research — UFZ
Permoserstraße 15
Leipzig, Germany

ISBN 978-94-007-2764-9 ISBN 978-94-007-2765-6 (eBook)
DOI 10.1007/978-94-007-2765-6
Springer Dordrecht Heidelberg New York London

Library of Congress Control Number: 2012945789

Printed on acid-free paper

Springer is part of Springer Science+Business Media (www.springer.com)

Acknowledgments

This study has been conducted during my stay at the Department of Economics at the Helmholtz Center for Environmental research from September 2007 to December 2010. I would like to take this opportunity to express my gratitude to the following people:

First and foremost, I would like to express my sincere gratitude to Prof. Dr. Bernd Hansjürgens for his supervision and guidance. With his kind help, I received the precious chance to do my Ph.D. study in UFZ. Also I have been receiving his continuous support during the entire time of my research stay. He provided lots of thorough and constructive suggestions on my dissertation. Secondly, I would like to thank Prof. Dr. -Ing. Rober Holländer for his willingness to supervise me, and his continuous support so that I can deliver my thesis at the University of Leipzig. Thirdly, I am heartily thankful to Dr. Nele Lienhoop, who helped me a lot in completing the writing of this dissertation. She was always there to meet and talk about my ideas and to ask me good questions to help me.

Furthermore, there are lots of other people who I would like to thank: Sara Herkle provided the survey data collected in Leipzig and Halle, Germany. Without these data, my thesis could not have been completed. It is my great honor to thank Prof. John B. Loomis (Department of Agriculture and Resources Economics, Colorado State University), who was always very patient to answer my questions and offered very good suggestions on the CVM study. I would like to thank those who provided valuable information and advice on conservation project of YFPs, including Secretary-General Prof. Nianhua Dai (Jiangxi Academy of Science, Ecological Society of Jiangxi Province, Nanchang, China), Prof. Ding Wang (Institute of Hydrobiology, Chinese Academy of Science, Wuhan, China), Associate Prof. Kexiong Wang (Institute of Hydrobiology, Chinese Academy of Science, Wuhan, China), Dr. Xiujiang Zhao (Institute of Hydrobiology, Chinese Academy of Science, Wuhan, China), Shuyuan Wu (Section chief of Resources and Environmental Protection, Jiangxi Administration of Fishery Bureau, Nanchang, China).

I offer my regards and blessings to all of those who supported me in any respect during the completion of the thesis, including Dr. Katharine N. Farrell (who gives me encouragement and confidence sustaining me overcome all the difficulties, helps me in proof-reading several chapters and statistics), Alexia Peterson (my International-Cultural Communication teacher, who taught me how to deal with business in a completely different cultural background. Her kind assistance made my life more comfortable and easier), Dr. Paul Lehmann (my office mate, who never hesitated to offer his kind help) and Aaron Leopold (my office mate, who helped me a lot with my English). It is also a great opportunity to show my great gratitude to three secretaries, Helga Dietsch, Monika Nussbaum, and Birgit Stumvoll who helped me a great deal with administrative matters and a pleasure to thank all the colleagues in Department ÖKUS and Chinese colleagues at the UFZ, who helped broaden my view and knowledge, and made lots of fun in my life. It is my great honor to meet and get to know you all in Leipzig.

Besides, I would like to thank the Helmholtz Center for Environmental Research-UFZ for providing a good environment and facilities to complete this project. The UFZ also financed my field trip to China. Also, I would like to take this opportunity to thank to the Junior Scientist Exchange Program between the China Scholarship Council (CSC) and the Helmholtz Association of German Research Centers for offering the financial support.

Finally, I am indebted to my lovely family: my parents, Hongmei Liu and Ruihe Dong, for giving me life in the first place, for educating me with aspects from both arts and sciences, for educating me how to be a useful person, for unconditional support and encouragement to pursue my interests, even when interests went beyond boundaries of language, field, and geography. My brother, Gang Dong and his wife, Yuanyuan Wang, for listening to my complaints and frustrations, and for believing in me, and for kind help in my life and work.

Contents

Abbreviations

ANOVA	Analysis of variance
ATP	Ability to pay
Asymp	Sig. Asymptotic significance
CAS	China Academy of Sciences
CS	Compensating surplus
CV	Compensating variation
CV	Contingent valuation
CVM	Contingent valuation method
DC	Dichotomous choice
df	Degrees of freedom
ES	Equivalent surplus
EV	Equivalent variation
GDP	Gross domestic product
HPM	Hedonic pricing method
IHB	Institute of Hydrobiology
IUCN	International Union for Conservation of Nature
MS	Market stall
MVPE	Market valuation of physical effects
NOAA	National Oceanic and Atmospheric Administration
RMB	Ren Min Bi, unit of Chinese currency
TCM	Travel cost method
TEV	Total economic value
UNEP	United Nations Environment Programme
WTA	Willingness to accept
WTP	Willingness to pay
WWF	World Wildlife Fund
YFP	Yangtze finless porpoise

List of Figures

List of Tables

Chapter 1
Introduction

Abstract Yangtze finless porpoises (YFPs) are the only fresh water adapted porpoises in the world, living in the middle and downstream of the Yangtze River, in China. They are facing the threats include: illegal fishing and overfishing, sand dredging and shipping, and water pollution caused by industry and agriculture. Owing to these threats, the porpoise population is falling and urgent conservation is needed. Poyang Lake is the largest freshwater lake in China and also the best habitat for YFPs. It is proposed a nature reserve for YFPs in Poyang Lake in order to avoid their extinction. The overall goal of the study is to quantify the economic benefits of the porpoise conservation in Poyang Lake in China using a deliberative valuation approach, called the valuation workshop. Within this overall goal, three sub-goals are pursued in the Poyang Lake valuation study, which seek to close specific gaps in the research on contingent valuation. Firstly, this study aims at finding a way that enables people to value an unfamiliar species, which mainly consists of non-use values. Secondly, it is to test empirically the role of the distance decay on WTP. Thirdly, the mean WTP for the porpoise conservation is compared between China and Germany.

Keywords Yangtze finless porpoises • Poyang Lake • Threats • Porpoise conservation • Economic benefits

1.1 Problem Statement

Yangtze finless porpoises (YFPs) are the only fresh water adapted porpoises in the world, living in the middle and downstream of the Yangtze River, in China. During the past decades, the population of YFP decreased greatly from about 2,700 to about 1,800 individuals due to a variety of threats. These threats include: illegal fishing and overfishing, sand dredging and shipping, and water pollution caused by industry and agriculture. Owing to these threats, the porpoise population is falling and urgent

Y. Dong, *Contingent Valuation of Yangtze Finless Porpoises in Poyang Lake, China*,
DOI 10.1007/978-94-007-2765-6_1,

conservation is needed. China is already considering upgrading the conservation status of YFPs, from a National II to National I conservation grade.

Poyang Lake is the largest freshwater lake in China and serves as an important habitat for YFPs. It is located in the northern part of Jiangxi Province and at the southern bank of the middle and lower reaches of the Yangtze River. It lies about 50 km north of the Nanchang city. At the same time, it is also called the last basin of clean water in China, with relatively better water quality compared to that of other lakes. Accordingly, it holds important biological sources and works as an important habitat for many species. Among these species, the YFPs are the one which is most prominent.

Poyang Lake is particularly suited for the YFPs at present for several reasons. First, there are about 400 porpoises living in Poyang Lake, accounting for about 25% of the total population of this species. Secondly, in the lake, the porpoise population is falling at a lower rate, compared with the other reaches of the Yangtze River system. Thirdly, the habitat of the YFPs is dramatically shrinking in other parts of the Yangtze River system. Finally, Poyang Lake holds great environmental capacity for sustaining the YFPs.

However, due to sand dredging, shipping, overfishing, and illegal fishing in Poyang Lake, the YFPs are not completely safe. Moreover, there is a hydropower and flood protection project in planning, involving the construction of several dams, which will fragment their habitat and cut off migration routines. The threats brought about by the Poyang Lake project are potentially serious. Hence, it is very urgent to take protection measures for the YFPs in Poyang Lake immediately. Therefore, a nature reserve is proposed by biologists in Poyang Lake in order to strengthen the conservation of the YFPs.

In order to find out whether a nature reserve is worthwhile, economic valuation can be a fruitful tool. Economic valuation is a means to estimate the values of certain species in monetary terms. It can help to make values visible and to mainstream ecological values in other policy fields. The Economics of Ecosystems and Biodiversity (TEEB 2010) is an impressive process demonstrating the aims and values of economic valuation in the field of ecosystem and biodiversity (see http://www.teebweb.org/).

As will be shown in the subsequent Chap. 2, among economic valuation methods, contingent valuation is in particular suited to evaluate the values of the species "YFP." The contingent valuation method (CVM) is used to estimate economic benefits of protecting the YFPs. It can be used to estimate both use and nonuse values, and it is the most widely used method for estimating nonuse values. The contingent valuation method involves directly asking people, in a survey, how much they would be willing to pay for specific environmental services. It is called "contingent" valuation, because people are asked to state their willingness to pay, *contingent* on a specific hypothetical scenario and description of the environmental service under investigation. In this study, a contingent valuation study is conducted. People are asked how much they would like to pay for a proposed nature reserve special for the YFPs in Poyang Lake.

1.2 Goal of the Study

The overall goal of the study is to quantify the economic benefits of the porpoise conservation in Poyang Lake in China using a deliberative valuation approach, called the valuation workshop. The study aims at eliciting thoughtful and informed values from a small number of people. That is, the intention is to reveal valid willingness to pay (WTP) estimates and calculate the mean WTP. Within this overall goal, three subgoals are pursued in the Poyang Lake valuation study, which seek to close specific gaps in the research on contingent valuation.

First, this study aims at finding a way that enables people to value an unfamiliar species, which mainly consists of nonuse values. It is likely that people have no preference for the YFPs ready in their mind. Since the main task of a CVM study is to produce valid WTP estimates for the porpoise conservation, a conventional survey method may not be suitable. A more elaborate approach is required, which can make respondents motivated and well-informed, and provides individuals enough time to consider about the WTP question. The valuation workshop method, one group-based approach, is selected to support respondents in constructing their preference. The advantages of the valuation workshop are detected by assessing the validity of WTP estimates.

Secondly, a subgoal of this study is to test empirically the role of the distance decay on WTP. So far, there are already some CVM studies on distance effects but there is no consistent conclusion. Distance effects work differently on different environmental goods and services in different case studies. The majority of existing CVM studies utilizes a predetermined geographic boundary for their sample frame (Pate and Loomis 1997). On the one hand, when distance is not incorporated into the analysis, it is possible that some individuals with positive values are not considered when too small a boundary is placed on the geographic extent of influence. On the other hand, similarly, values may be overstated if the sample estimate is extrapolated over too large a region (Sutherland and Walsh 1985). Therefore, there is a need to establish the market extent of the values of the YFPs by testing the distance decay on WTP, before aggregating benefits estimates.

Thirdly, the mean WTP for the porpoise conservation is compared between China and Germany. In general, present research only concentrates on investigating people's preference on certain environmental good in a local, state, or nationwide range. However, the YFPs are unique endangered rare mammals. The specific objectives of the international comparison of WTP for the porpoise conservation are to:

- Detect whether income effects on WTP still work in a worldwide range. The YFPs live in China and it is expected that people in China should care more about them than people in Germany. However, Germans have a higher average income than Chinese people. Therefore, it is meaningful to check whether Germans would like to contribute more to the project than Chinese people, due to their higher income.

- Test whether budget constraints influence individuals' WTP. Is it possible that Chinese stated WTP will be distorted by their ability to pay?
- Establish the market extent of the YFPs mainly driven by nonuse values.

Thus, the goal of this study contributes to closing methodological gaps in recent research on CV, and to contribute empirically to a pressing issue in China's environmental policy—the protection of a highly visible endangered species.

1.3 Structure of the Thesis

The thesis is organized as follows. In Chap. 2, the background information of Poyang Lake and the YFPs will be provided. Chapter 3 introduces the theoretical background underlying economic valuation and reviews different valuation methods. In this chapter, the theoretical basis for CVM will be explained, including welfare measures, WTP, and WTA. Moreover, the challenges faced with CVM will be summarized and discussed. In Chap. 4, the questionnaire development, including information folder, hypothetical market, some socioeconomic questions as well as general questions on environmental issues—an important share of a CVM study—will be illustrated. Furthermore, different survey methods will be compared in the light of value elicitation for the YFPs. The results of the empirical analysis will be presented in Chap. 5, including a summary of sample characteristics, descriptions of individual attitudes toward the porpoise conservation, and mean WTP values. Chapters 6 and 7 discuss the implications of the research and draw some conclusions on the performance of CVM, the distance effects, and the preferences of Chinese people and Germans. Finally, some suggestions for future research will be put forward.

References

Pate, J., & Loomis, J. (1997). The effect of distance on willingness to pay values: A case study of wetland and salmon in California. *Ecological Economics, 20*, 199–207.

Sutherland, R. J., & Walsh, R. (1985). Effect of distance on the preservation value of water quality. *Land Economics, 61*, 281–291.

The Economics of Ecosystems and Biodiversity (TEEB). (2010). http://www.teebweb.org/

Chapter 2
Background Information of Poyang Lake and Yangtze Finless Porpoises

Abstract Before I start describing Yangtze finless porpoises (YFPs), I need to put this animal and the need to protect it into context. Therefore, this chapter seeks to introduce the area, Poyang Lake, and identify major use conflicts with respect to ecology and economy. In addition, I briefly describe the characteristics of the YFPs and explain the threats faced by them and corresponding protection strategy.

Poyang Lake enjoys the title of the last "special water ecological region" in the world, gaining the laureate of the "largest freshwater lake of China" for its maximum water surface of 5,100 km² in 1954. Poyang Lake is not only the lifeline of Jiangxi province in China but also the "patron saint" of the middle and lower reaches of The Yangtze River. In this chapter, I will introduce the geographical situation of Poyang Lake, and the uses and conflicts found there (Sects. 2.1, 2.2, and 2.3). Then I will explore Poyang Lake as the best currently available habitat of the YFPs and explain the circumstances faced by the YFPs (Sect. 2.4). The chapter ends with a brief summary (Sect. 2.5).

Keywords Yangtze finless porpoises • Poyang Lake • Threats • Conservation • Nature reserve

2.1 Introduction to Poyang Lake

2.1.1 Geographical Situation

Poyang Lake is located at latitude 28°22′–29°45′ north and longitude 115°47′–116°45′ east. It lies in the northern part of Jiangxi Province, at the southern bank of the middle and lower reaches of the Yangtze River (see Fig. 2.1). It is divided into two parts by the Songmenshan Mountain (Wu 2008). The northern part is the water channel joining the Yangtze River, with a length of 40 km and a width of 3–5 km (the narrowest point is 2.8 km). The southern part is the main lake, with a length

Y. Dong, *Contingent Valuation of Yangtze Finless Porpoises in Poyang Lake, China*,
DOI 10.1007/978-94-007-2765-6_2, © Springer Science+Business Media Dordrecht 2013

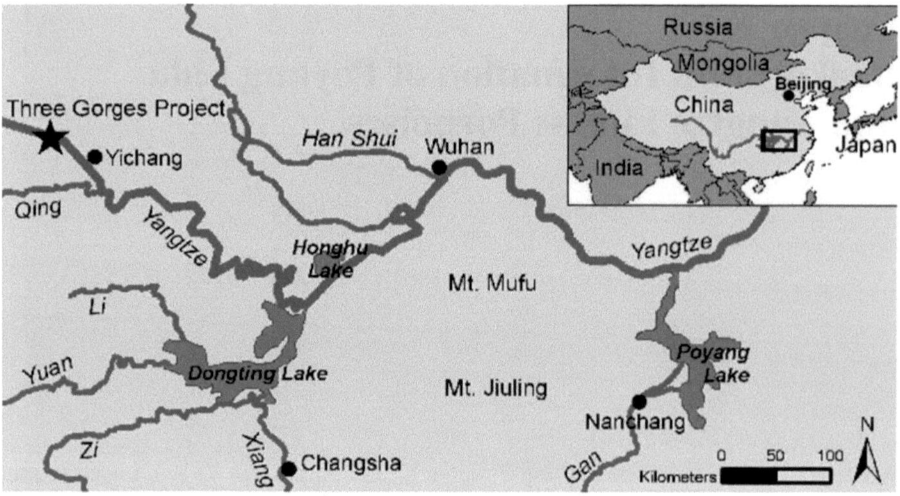

Fig. 2.1 Location of Poyang Lake in China (Source: Te Boekhorst et al. 2010)

of 173 km and its furthest width of 70 km with a mean width of 16.9 km from west to east (Xu et al. 2001).

2.1.2 The Largest Freshwater Lake in China

Poyang Lake is the largest fresh lake in China. The lakeshore is 1,200 km long, and the lake surface area of the water body can range from 4,070 km^2 (when the water level at Hukou station was 22.59 m in 1998) to 146 km^2 at the water level of 5.90 m in 1963 (Liu 2003). Its volume is 27.6 billion m^3. It retains water from the five rivers—Ganjiang, Fuhe, Xinhe, Raohe, and Xiuhe, and empties into the Yangtze River, the largest and most important river in China, at Hukou station after storing up its water (see Fig. 2.2). The Ganjiang River is the largest river in the region, extending 750 km. It contributes 55 % of the total discharge into Poyang Lake (Shankman and Liang 2003). Poyang is a seasonal lake with the feature of taking in and sending out water. The rainy season in Jiangxi usually begins in April. Most years, the five rivers' discharge increases from April to June, raising the level of Poyang Lake, which drain into the Yangtze River. From July to September, the discharge from the five river decreases. But at the same time, the water level of the Yangtze River increases. As a result, usually mid-July, the water flow from the lake into the Yangtze River reverses and begins moving from the Yangtze River into Poyang Lake. Maximum Yangtze River discharge typically occurs during the mid- to late-summer months. The area of the watershed is 162,000 km^2 (Shankman and Liang 2003). The annual mean runoff of the watershed is 149.4 billion m^3, accounting for 16.7 % of that of the watershed of the Yangtze River (Xu et al. 2001).

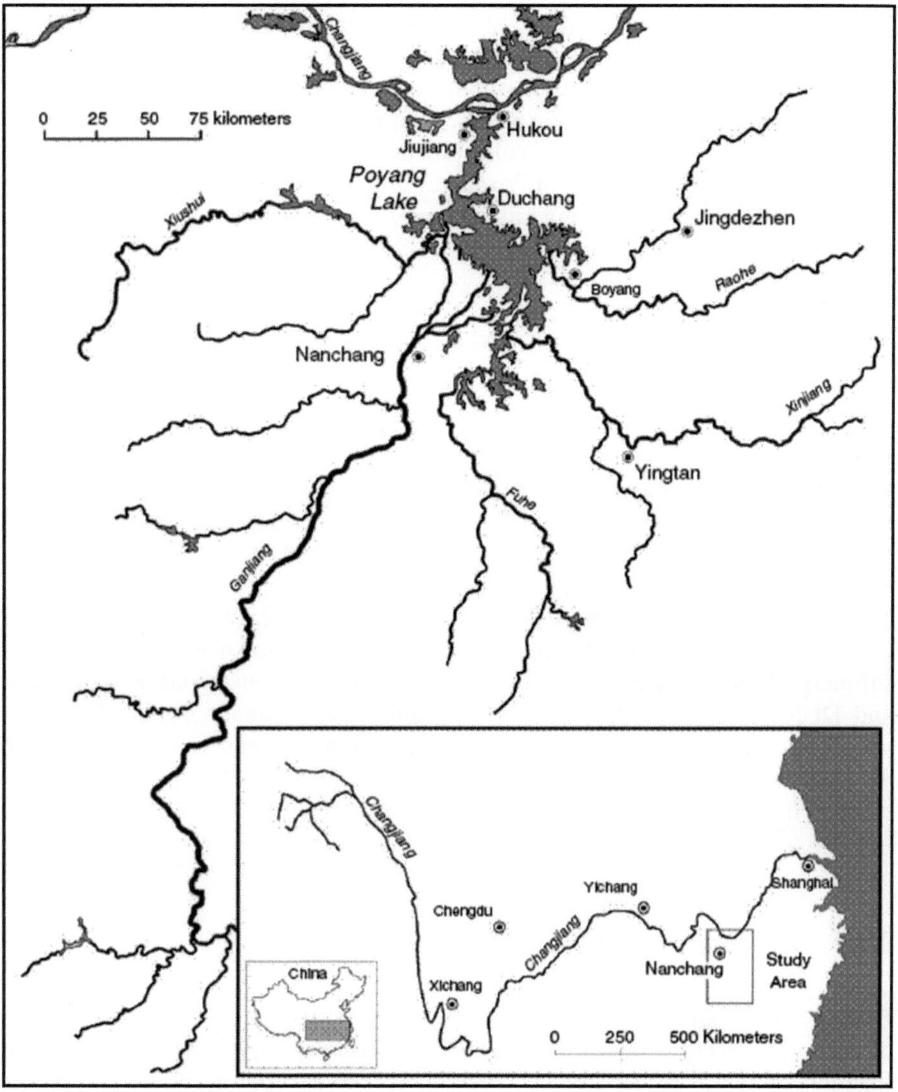

Fig. 2.2 Map of Poyang Lake and major rivers in Jiangxi Province (Source: Shankman et al. 2006)

2.1.3 Landscape—"A Line at Low Water, An Ocean at Flood"

The mean water level of Poyang Lake in many years is 12.86 m, the highest was 22.59 m on July 31, 1998, the lowest was 5.9 m on February 6, 1963 (at Hukou Hydrological Station, see Fig. 2.2). The average amplitude of variation of water level is from 9.79 to 15.36 m in a year. The absolute fluctuation of water level is up to

16.69 m (see: Overview of Poyang Lake). With variation of precipitation in different seasons, the fluctuating range of water level is relatively large. Correspondingly, the area of Poyang Lake greatly varies with the fluctuation of its water level. The water level rises during the rainy season and then the water surface expands. And it drops during the dry season and bottomland comes out, and only several wandering watercourses remain. The landscape of Poyang Lake looks like a line during dry seasons and it looks like an ocean during rainy seasons.

2.1.4 Evolutional History

In ancient periods, Poyang Lake was called Pengli Pond, Guantinghu Lake, and many other names. Over a very long time, Penglize Bog expanded to the south, and water went over the Songmenshan Mountain and reached over the vicinity of Poyang County—through the interaction of geological, meteorological, and hydrological factors. As a result, its name was changed to Poyang Lake. Before the invasion of water to the south, the south to Songmenshan Mountain was originally a well-populated Xiaoyang Plain. With gradual expansion of water to the south, Xiaoyang County and Haihun County in the basin of Poyang Lake were successively submerged under water. There is a saying that "Xiaoyang county submerged and Duchang county appeared, and Haihun county submerged and Wucheng county appeared." Through this long-term evolution, the modern miniature of the mist-covered and vast Poyang Lake formed about 1,600 years ago (see: History of Evolution).

2.1.5 An Important International Wetland

Poyang Lake is an important international wetland, and an important storing lake of the main stream of the Yangtze River. It possesses very important ecological functions, e.g., floodwater storage and biological diversity protection, in the watershed of the Yangtze River. Poyang Lake is the treasure land for flood regulation and storage of the Yangtze River (Zhu et al. 2004; Cui 2004). It is one of the ten ecological conservation areas in China, and also one of the global important ecological areas regulated by Global Environment Facility (Secretariat of 11th Living Lakes Conference 2006). It plays a very important role in maintaining the ecological vitality of the region and even the whole country.

So far, we have gained a general impression of the geographical and hydrological characteristics, evolutional history, and important status of Poyang Lake. In the following section, I will turn to consider the current uses of Poyang Lake.

2.2 Uses of Poyang Lake

Poyang Lake is the mother lake of Jiangxi Province, which support a population of around ten million people in the vicinity of the lake. At the same time, there are abundant biological resources, such as aquatic plant, fishes, birds, and the YFPs. Fresh water and plentiful resources give the Poyang Lake its important role.

Poyang Lake contributes many services to humans. Among those services are ecological ones as well as economic and social ones. Important economic and social uses of Poyang Lake include industrial water use, agricultural water use, fishery, and ecotourism. As for the ecological uses of Poyang Lake, it is the important habitat for migratory birds, fish resources, and the YFPs.

2.2.1 Industrial Water Use

There are ten main industries, including manufacture of autos, air conditioner, food, photo-electricity and software, paper making, airplane manufacture, fabric spinning and clothes making, manufacture of medicine and medical appliance, of electronic mechanisms, and of building materials and other new materials. In 2003, the annual *industrial water consumption* was 1.593 billion m^3, accounting for 22.6 % of total water consumption in Poyang Lake region (Xu 2005). With the rapid development of economy, more and more water is needed for industrial use. From 2003 to 2007, there was a stress situation with the industrial water supply (Sun 2009). People are facing water scarcity in the Poyang Lake region. In 2005, there was water shortage of 0.488 billion m^3, accounting for 11 % of the total water requirement (Wang 2008).

2.2.2 Agricultural Water Use

From 2003 to 2008, the average annual water consumption was 4.272 billion m^3 in the Poyang Lake region, while the annual agricultural water consumption was 2.98 billion m^3, accounting for 69.9 % of total consumption (Wang 2008). There are good conditions favorable for agricultural production in the vicinity of the lake. However, precipitation is distributed unevenly over time and space. From July to September when temperature and evaporation is high, there is little precipitation, which is not good for the growth of late rice. In order to meet crops' water demands, a great number of irrigation projects have been initiated. There are about 2,270 water-storing projects (including small reservoirs and ponds on the hills), with the capacity of 10.906 billion m^3 and the effective capacity of 5.446 billion m^3. Among them, there are six large reservoirs with a combined total capacity of 8.709 billion m^3

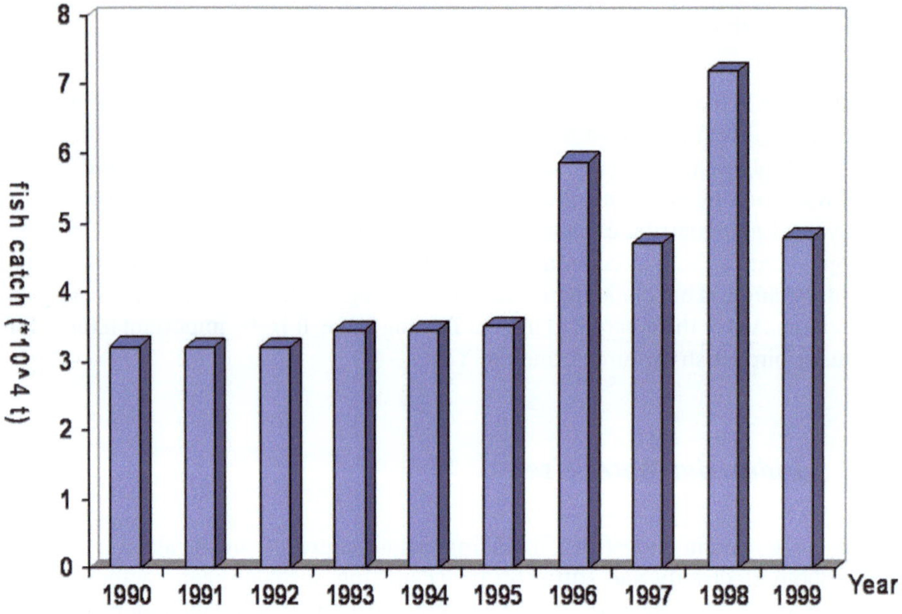

Fig. 2.3 Fish catch in Poyang Lake in 1990s (Data source: Huang and Huang 2007)

and an effective capacity of 3.875 billion m^3, and there are 39 medium reservoirs with a combined total capacity of 846 million m^3 and an effective capacity of 573 million m^3. There are 4,880 irrigation pumping projects with a total installed pumping capacity of 240,000 kW-hours per year. There is an effective irrigation area of 4,767 km^2 in total with an area of 3,843 km^2 promising stable yields regardless of drought or excessive rain, respectively accounting for 83.6 and 67.4 % of the total arable land area in Jiangxi Province (*Honkong Business Newspaper* 2009).

2.2.3 Fisheries

The production of freshwater fish from Poyang Lake accounts for 17.50 % of the total production in China, 46.67 % of that coming from the Yangtze River, and 68.29 % of the production within Jiangxi province. About 100,000 fishermen live around the Poyang Lake. However, the lake has suffered a decline in fish reproduction for two decades, since 1980s, as a result of overfishing (see Fig. 2.3), pollution, and depletion of sand in the lake area (Xinhua News Agency 2010). In order to ensure the sustainable development of the local fishing industry, the policy of a Spring Fishing Ban of 3 months has been implemented in Poyang Lake since 2002. The policy is a success as some species have reappeared in the lake after years of absence and fishermen's income increased when fishing resumed in summer (Xinhua News Agency 2009).

2.2.4 Ecotourism

There are several opportunities for the development of ecotourism in the Poyang Lake region. In autumn and winter, thousands of rare fowl and migratory birds fly to Poyang Lake for the winter, forming a unique landscape of migratory birds.

There are ongoing plans to use the international popularity and reputation as "World Well-known Wetlands," "Rare Bird Kingdom," "Migratory Birds Paradise" to promote migratory birds' ecotourism and wetland adventure ecotourism, on the basis of comprehensively exploiting grass, tidal flats, aquaculture, and so on (China Central TV 2010). Since the beginning of 2010, ecotourism has become the preferential policy of governments in many areas around Poyang Lake. Also, it is possible to develop the sightseeing of the YFPs. For example, the establishment of watching docks for the YFPs in Poyang Lake is under consideration.

2.2.5 "The World of White Cranes" and "The Kingdom of Rare Birds"

Since Poyang Lake possesses a humid monsoon climate, Poyang Lake has become, accordingly, "the country of rivers and green grass, and blooming plums in the misty rain," and "a land flowing with milk and honey." Its environment and climatic conditions are suitable for migratory birds to live through winter. At late autumn and early winter in every year, thousands and thousands of birds migrate over here from Siberia of Russia, Mongolia, Japan, Korea, and northeastern and northwestern China. At spring (April) of the following year, they gradually migrate away (Bird Life International 2010).

Poyang Lake accommodates 332 kinds of birds, subordinate to 17 orders and 55 families. There are 124 species of waterfowls: 32 species of ducks (*Anatidae*) and 25 species of snipes (*Scolopacidae*). Ten of them are first class species on China's list of protected species ("Red List"), among them the famous Siberian Crane (*Grus leucogeranus*) is the symbol of Poyang Lake. Almost 95 % of Siberian Cranes in the world spend their winter in Poyang Lake wetlands. Other rare and endangered species wintering at the lake include the Oriental White Stork (*Ciconia boyciana*), the Black Stork (*Ciconia nigra*), Swan Goose (*Anser cygnoides*), White-naped Crane (*Grus vipio*), Hooded Crane (*Grus monacha*), and the Great Bustard (*Otis tarda*). Forty-four of the wintering species are on the second national list of protected birds, including the Sacred Ibis (*Threskiornis aethiopica*), the Spoonbill (*Platalea leucorodia*), and the White fronted goose (*Anser albifrons*) (Convention on

Migratory Species). This is why Poyang Lake came to be known as "the world of white cranes" and "the kingdom of rare birds."

2.2.6 *Fish Source*

Poyang Lake belongs to the humid subtropical monsoon climate, with a warm climate, adequate sunlight, ample rainfall, and a long frost-free period, while being hot and humid at the same time. These water and weather conditions, accompanied by the broad lake and grassland, are ideal for fish spawning and propagation for species such as crucian carps. It is also suitable for migratory fish and semimigratory fish, which spawn and spend a period of juvenile growth in Poyang Lake. It is recorded that there are 136 fish species belonging to 25 families and 78 genera. The largest family is *Cyprinidae* (71 species), accounting for up to 52.2 % (Zhang and Li 2007). Poyang Lake is rich in aquatic plants, which create a hospitable environment for many rare species of freshwater fish, such as *Leiocassis Longirostis* and the Chinese Sucker, which are both rare fish species, as rated by the State (Chen et al. 2004; Zhang and Li 2007). Plentiful fish resources play an important role in maintaining the biodiversity of Poyang Lake region, and potentially in a wider range of ecosystems.

2.2.7 *The Best Habitat of Yangtze Finless Porpoises*

Poyang Lake is regarded at present as the best habitat for the YFPs on several grounds. First, there are around 400 YFPs living in Poyang Lake, while there are 1,800 YFPs living in China (Xiao and Zhang 2000; Zhao et al. 2008). The porpoise population proportion is great in Poyang Lake, almost 25 % of the total population. Secondly, the rate of decrease of the porpoise population in Poyang Lake is the lowest in China, as compared with that in other parts of the Yangtze River system. The population of the YFPs in the whole Yangtze River system is declining at the rate of 5 % per year and the porpoise population is decreasing at the exponential rate of 13 % per year in the Balijiang section of 40 km, which is the confluence area of Poyang Lake and the Yangtze River (Wei et al. 2002; Zhao et al. 2008). Thirdly, many female porpoises are pregnant in Poyang Lake, according to the latest survey results. But it seems that there are more female YFPs than male porpoises in Poyang Lake. It is necessary that some male porpoises are introduced into Poyang Lake. Fourthly, the density of YFPs is particularly high in Poyang Lake (Wang et al. 2000). The observed density is only 0.02 porpoise/km in the 716.4 km section of the Yangtze River, stretching from Yichang to Ezhou, and the YFPs appear to be at the highest risk of local extinction in this section. No YFP was detected in the subsection of 150 km from Shishou to Yueyang, which indicated that the porpoise population must be very few (Zhao et al. 2008). All these facts tell that the habitat of YFPs is shrinking greatly, which emphasizes the important status of Poyang Lake as one of the few remaining habitats of the YFPs. Finally, it is also estimated that the environmental capacity of Poyang Lake to support a larger porpoise population is potentially great. If the ecology can be well-maintained, Poyang Lake could hold up to 1,000 porpoises.

2.3 Use Conflicts in Poyang Lake

From the above description of water uses, it becomes clear that a number of water uses are producing conflicts in Poyang Lake. The main water use conflicts in Poyang Lake are around sand dredging, building dams, water shortage, and water pollution. Among them, sand dredging, as the mainstay of local economic development, is an important source of fiscal revenue in the region that borders Poyang Lake. However, at the same time, high-density dredging projects have been the principal cause of the death of the local wildlife population.

2.3.1 Conflicts Resulting from Sand Mining

Recently, intensive sand mining has been reported in Poyang Lake. Zhong and Chen (2005) attributed the intensive sand mining in Poyang Lake to a few factors. First, the demand for sand is driven by fast economic development in the lower Yangtze River basin, where mega cities like Shanghai and Wuhan are rapidly expanding. Secondly, when sand dredging was banned in the middle and lower Yangtze River in 2000 (*People's Daily* 2000), it then developed prosperously in Poyang Lake. Thirdly, with the price of sand and gravel increasing, more and more sand mining boats come out onto Poyang Lake. At present, there are 160 sand dredging enterprises operating on the lake. In total, there are hundreds of sand dredging boats and ships. It is recorded that in 2005 profits of more 25 million RMB were gained for the local economy from sand dredging. At peak times, there are more than 200 tug boats mining sand on the lake. The power of most sand dredgers is above 3,000 kW, each allowing a per hour mining capacity of 10,000 t approximately (Li 2008).

Sand dredging impacts on *water quality,* increasing turbidity, and freeing up heavy metals, nutrients, pesticides, and PCB's contents from the lake bed (Groot 1979). Regardless of what method of mining is selected, an increase in the turbidity in the area will always occur. The effects of turbidity include that light penetration is reduced and the lake bed is covered with dredging discharges. As a result, sand mining produces negative impact on the *biodiversity* of Poyang Lake. Water turbidity produced by sand dredging reduces light intensity and therefore also the productivity of *Vallisneria spiralis* (Wu et al. 2007). The tubers of *Vallisneria spiralis* form an important food source of the endangered Siberian cranes (Kanai et al. 2002), which winter almost exclusively in Poyang Lake (De Leeuw et al. 2009).

Sand mining also threats YFPs due to noise pollution, accidental entanglement, reduction in their food sources and chemical water pollution (Wang et al. 2005).

China has banned river sand mining in Poyang Lake since April 1, 2008 in order to protect its aquatic environment. Sand mining will resume in designated areas with limits on the output after an official plan on the mining comes out, although no time frame for the mapping plan was given (Xinhua News Agency 2008). However, the environmental impact caused by sand mining still exists in Poyang Lake. Besides the sand mining, another serious water use conflict is the building of levees and dams in Poyang Lake, which is explained in the next section.

2.3.2 Conflicts from Building Levees

For centuries, levees in Jiangxi province were used to protect cities and farmland. However, during the late 1800s and early 1900s, war and floods destroyed many of the levees. Before 1950, the total length of levees in Jiangxi Province was about 3,100 km. Since the early 1950s, the Jiangxi Provincial Government has conducted major levee construction projects and there are now about 6,400 km of levees (Peng 1999; Shankman and Liang 2003).

Now there is a Poyang Lake project put forward by Water Resources Department of Jiangxi Province and Jiangxi Provincial Development and Reform Commission. The proposal of the project was submitted on January 14, 2009. The Poyang Lake project is a large-sized hydro complex, which aims to protect both life and health security of local people and Poyang Lake's ecoenvironment through a comprehensive development and utilization of water resources in the Lake District. Poyang Lake project consists of several parts, including Poyang Lake dam, hydropower plant, flood diversion sluice, navigation lock, fish passage, and rafting passage (see Fig. 2.4) (see: Jiangxi Water Conservancy 2010).

Levees increase lake-sediment deposition and, thus, reduce the lake volume. As a result, the frequency and magnitude of floods in Poyang Lake region are increasing (Shankman and Liang 2003). With the damming of Poyang Lake, it is highly likely that the flora and fauna as well as the overall ecosystem function at Poyang Lake will suffer decline, which will have deleterious effects that extend beyond wildlife, and influence all users of the Poyang Lake system, particularly local human communities that depend on the lake for a variety of goods and services (Barzen et al. 2009). In addition, the damming might disrupt the migration route of the YFPs. Li et al. (2002) concluded that after the building up of Tongling bridge, the density of the YFPs decreased greatly in the water area close to the bridge, because the bridge piers caused a change in the shape of the river bed and in river flow. If the encountered rate of the YFPs decreased greatly and the migration of the YFPs was blocked because of a bridge, the same result might be expected from the damming of Poyang Lake. The damming of Poyang Lake is intended to reduce the natural variation of water levels within the system, and thereby to eliminate the water conflicts here. Now I turn to the conflicts with respect to water quantity and quality.

Fig. 2.4 Schematic of the Poyang Lake Project (Source: Jiangxi Water Conservancy 2010)

2.3.3 Conflicts with Respect to Water Quantity and Quality

As mentioned in the section above, on industrial water use, there is water scarcity in Poyang Lake region. In the face of this water scarcity, some conflicts have arisen, both with regard to water quantity and water quality. That is, less water is supplied than is needed and some water is of bad quality, so it cannot be used.

In Poyang Lake, water shortage can be seen in the continuous low water levels and drought hazards. The Poyang Lake system has, in recent years, consistently been breaking the records of unusual historical low water levels, and with prolonged durations. In 2007, there was an unusual historical low water levels with a duration of more than 100 days. Drought disasters are serious in Poyang Lake region and even in Jiangxi province. In 2007, the area of Poyang Lake dwindled to 50 km^2, less than 1 % of the largest lake area with water volume of only 12 million m^3 at one time. As a result, 11 cities were more or less suffering ordeals of inadequate drinking water. There were 760,000 people and 310,000 big animals lacking drinking water. Agriculture also suffered a great loss. It is estimated that 3,521 km^2 of crops suffered drought and 640.32 km^2 of winter crops failed to be cultivated because of drought across Jiangxi province (see *Honkong Business Newspaper* 2009). During the drought period, some water conflicts arose between industrial water use, agricultural water use, and ecological water use. For instance, the ecological water use is at a very low level due to industrial and agricultural water overuse. As a result, flora and fauna in the lake were reduced greatly during the 2007 drought, which resulted in the shortages of food for fish and migratory birds, and also for the YFPs.

The rate of degradation of water quality has accelerated in recent years. Poyang Lake once had clean waters. In 2001, the water was classified as Category I or II on 80 % of tests; it was Category III in the remaining 20 %. These were the top three classes in China's scale of water quality, which meant that the water in Poyang Lake was suitable to drink all year round. By September 2007, however, a provincial monitoring center found no Category I or II water in ten sampled locations in the lake. (The water was at its normal level at that time and covered 3,000 km^2). The water was Category III in 60 % of the test areas in Poyang Lake. The remaining 40 % were Category IV: lightly polluted water. In December 2007 and January 2008—when the water level was very low and the lake covered only 40 km^2—monitoring indicated that the water flowing from the lake into the Yangtze River was Category V: heavily polluted water (Li 2008).

Exacerbating water pollution causes some serious environmental effects. For example, water pollution can drive many endangered rare species into extinction. Much wastewater, containing nutrients like phosphorus, is discharged into Poyang Lake, which speeds the reproduction of algae in the lake, throwing the ecosystem out of balance. When algae die and decompose, they reduce the oxygen level in water, so these pollution fostered booms can kill fish and other aquatic organisms. In addition, the algae reduce the amount of light penetrating the water, which can constrain the growth of other aquatic plants (Min 1998). At the same time, many other animals feeding on fish and aquatic plants, such as migratory birds and YFPs, are also negatively affected because of the impact that algae growth has on aquatic plants. In the end, biodiversity and ecological function of the lake system will deteriorate if this pollution continues.

In summary, sand mining, dams, water quantity, and water quality problems lead to main water conflicts in Poyang Lake. These conflicts are already or will soon pose serious threats to the survival of many rare species that depend on the lake, such as migratory birds and the YFPs. A great deal of research has been done on the protection of migratory birds, organized by International Crane Foundation and Poyang Lake Nature Reserve (see: International Crane Foundation, retrieved 2011). However, much less attention has been paid to the YFPs in Poyang Lake. This is one of the reasons that this study focuses on the protection of the YFPs in detail. Now I turn, in the following sections, to explain what the YFPs are, where they live, what threats they face, and why and how to protect them.

2.4 Yangtze Finless Porpoises

The Latin name of the finless porpoise is *Neophocaena phocaenoides* (see Fig. 2.5). The finless porpoise is small and lacks a dorsal fin. The fin is replaced by a ridge which runs down the middle of the back. The head is round and there is no apparent beak. The color is uniformly dark, ranging to pale grey and somewhat lighter on the ventral side. Body size normally reaches around 170 cm and weight around 70 kg, with respective maximums of around 200 cm and 100 kg (Amano 2002).

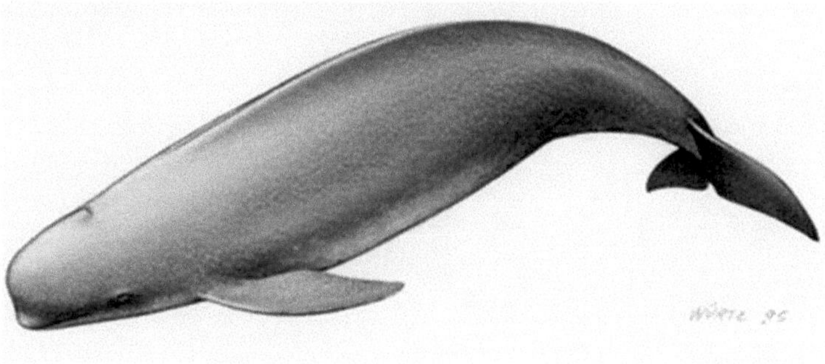

Fig. 2.5 Drawing of *Neophocaena phocaenoides* (Source: Würtz-Artescienza 2002)

2.4.1 Distribution and Category of Finless Porpoises

The warm, coastal Indo-Pacific waters, both fresh and marine, are home to the finless porpoise (see Fig. 2.6) (Jefferson and Hung 2004). Specifically, finless porpoises live in the coastal waters and the mouths of all the major rivers of the Indian Ocean and the Western Pacific.

There are three well-marked regional populations, which warrant subspecific ranking, while even within these three regional populations, significant differences in skull morphology have been found among local populations (Rice 1998; Amano 2002):

N. p. phocaenoides (also called the tropical finless porpoise) inhabits coastal waters along the mainland of southern Asia from the Persian Gulf east to the South China Sea and the southern part of the East China Sea; as well as to the coasts of south-eastern Sumatra, Bangka, Belitung, Sarawak, Palawan, the Turtle Islands in the Sulu Sea, and northern Java. The species has not been found in South African waters, or anywhere else in Africa (Rice 1998). It penetrates into the Indus River for 60 km, and into the Brahmaputra River for 40 km from the mouth (Kasuya 1999).

N. p. sunameri (Pilleri and Gihr 1972) (also called the temperate finless porpoise) ranges in the coastal waters from the southern East China Sea, north to the Liaodong Wan in China, to Korea, and to Kyushu in Japan, and from there along the Pacific coast of Japan from the Seto-naikai north to Sendai-wan in northern Honshu (Rice 1998). Five local populations are identified in Japanese waters based on skull morphology and mtDNA variability (Yoshida 2002; Amano 2002).

N. p. asiaeorientalis (Pilleri and Gihr 1972) is found in the lower and middle reaches of the Yangtze River, where it ranges 1,600 km upstream as far as the gorges above Yichang (200 m above sea level), and including the Poyang Lake and Dongting Lake and their tributaries, such as the Gan Jiang and the Xin Jiang (see Fig. 2.7) (Rice 1998). They are also called YFPs and are the species upon which this dissertation is concentrated.

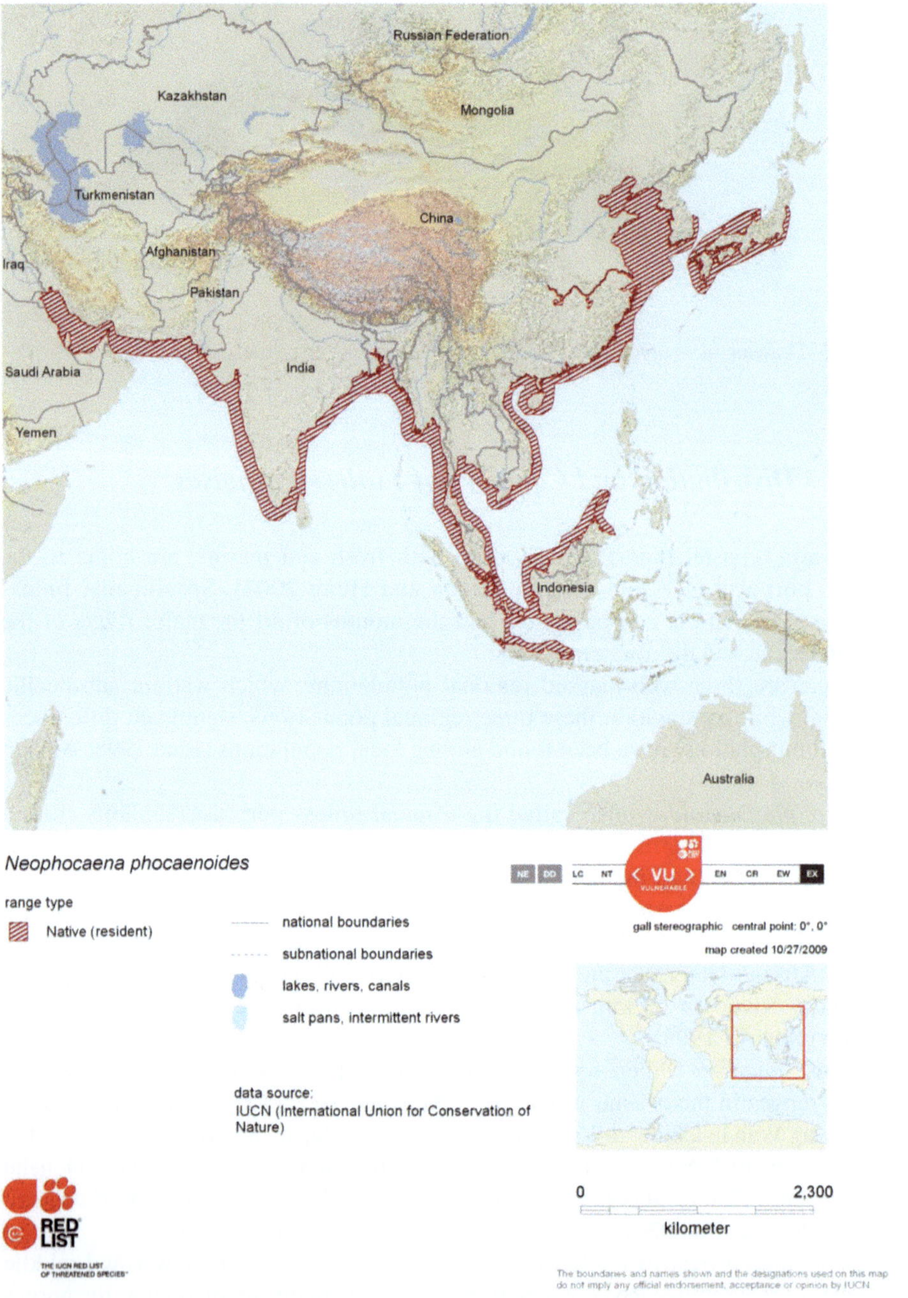

Fig. 2.6 Distribution of finless porpoises (Source: Reeves et al. 1997)

Fig. 2.7 Distribution of Yangtze finless porpoises (Source: Travel China Guide (2011), plus own presentation)

2.4.2 Biology and Behavior

2.4.2.1 Habitat

The finless porpoise is mainly an inshore species, but occurs in salt and fresh water. *N. Phocaenoides* appears to prefer murky or turbid conditions and can be found in warm rivers, lakes (if connected to rivers), mangroves, estuaries, deltas, and salt marshes. It prefers areas where river and ocean waters meet (Carwardine 1995). In the Yangtze River, the YFPs are found up to 1,600 km from the sea, whereas in Japanese waters they prefer shallow depths (<50 m) and close proximity to the shore (<5 km). In the shallow East China Sea, however, proximity to the shore is not so important (Shirakihara et al. 1992; Amano 2002).

2.4.2.2 Behavior

Like other porpoises, their behavior tends to be not as energetic and showy as that of dolphins. They do not ride bow waves, and in some areas appear to be shy of boats. Mothers have been seen carrying calves on the denticulate area on their backs.

In the Yangtze River, the YFPs are known to leap from the water and perform "tail stands" (Jefferson et al. 1993).

2.4.2.3 Schooling

Finless porpoises are generally found as singles, pairs, or in groups of up to 12, though aggregations of up to about 50 have been reported (Jefferson et al. 1993). Recent data suggest that the basic unit of a finless porpoise school is a mother/calf pair or two adults, and that schools of three or more individuals are aggregations of these units or of solitary individuals. Social structure seems to be underdeveloped in the species, and the mother/calf pair is probably the only stable social unit (Kasuya 1999).

2.4.2.4 Reproduction

Reproduction in most areas has not been well-studied. Reports indicate that calving in the Yangtze River occurs between April and May, whereas on the Pacific coast of Japan it occurs between May and June and in western Kyushu between November and December. Animals form Kyushu live around 25 years and attain sexual maturity at the age of between 4 and 9 years old. Gestation lasts 11 months (Amano 2002).

2.4.2.5 Food

Finless porpoises are reported to eat fish and shrimp in the Yangtze River; and fish, shrimp, and squid in the Yellow Sea/Bohai area and off the coast of Pakistan. In Japanese waters, they are known to eat fish, shrimp, squid, cuttle fish, and octopus. Finless porpoises are opportunistic feeders utilizing various kinds of food items available in their habitats. Seasonal changes in the diet have not been studied (Kasuya 1999), however, they also apparently ingest some plant material, including leaves and rice (Jefferson et al. 1993).

2.4.3 Migration

Available information suggests that the YFPs are probably found year-round throughout their range, and show various degrees of seasonal movement and density change, which are not well-documented in most areas (Kasuya 1999).

An annual migration is reported in the Inland Sea of Japan, where porpoises are faced with drastic seasonal changes in surface water temperature between 6 °C (March) and 28 °C (September). Their density is lowest (40 % of the peak season) in early winter, and starts to increase in January, reaching its peak in April. Finless porpoises migrate to and from the Pacific coast of Japan mainly through two passes

Fig. 2.8 Map of the Yangtze River (Source: Vocations To Go 2011)

at the eastern Inland Sea of Japan. From observations in the fluctuation of the proportion of mother–calf pairs, it is suggested that porpoises use the Inland Sea of Japan as a breeding ground. In summer, the animals move out to the Pacific coast (Reyes 1991; Kasuya 1999). Finless porpoises are known to be present year-round in Ise Bay and Mikawa Bay of Japan with peak abundance in April to June. They also can be found year-round off the coast of western Kyushu, where density is high in both winter and spring in coastal waters of less than 50 m depth (Culik 2004).

Along the Chinese coast, finless porpoises are present all year round, but some seasonal density changes are reported in the Bo Sea and on the Yellow Sea coast (see Fig. 2.8) (low in winter and high in summer/autumn). Therefore, they apparently move from shallow water to deeper water in winter. Movement of finless porpoises between the Yangtze and the ocean has yet to be confirmed (Kasuya 1999). Akamatsu et al. (2002) recently documented daily horizontal travel distances of two finless porpoises in the Yangtze River as 94.4 and 90.3 km.

In the Indus delta, finless porpoises move to the sea in April and return to the creeks and delta in October; here the movements of porpoises are said to follow movements of prawns (Reyes 1991).

Parsons (1998a) reports on a stranding of 154 small cetaceans in Hong Kong territorial waters with finless porpoises and Indo-Pacific hump-backed dolphins accounting for 77 % of these strandings of *N. phocaenoides*, which were more frequent in the winter. Almost a third of all finless porpoises stranded were calves. At sea finless porpoises were only sighted to the south of Lantau Island and were more frequently observed during winter. Their abundance was negatively correlated

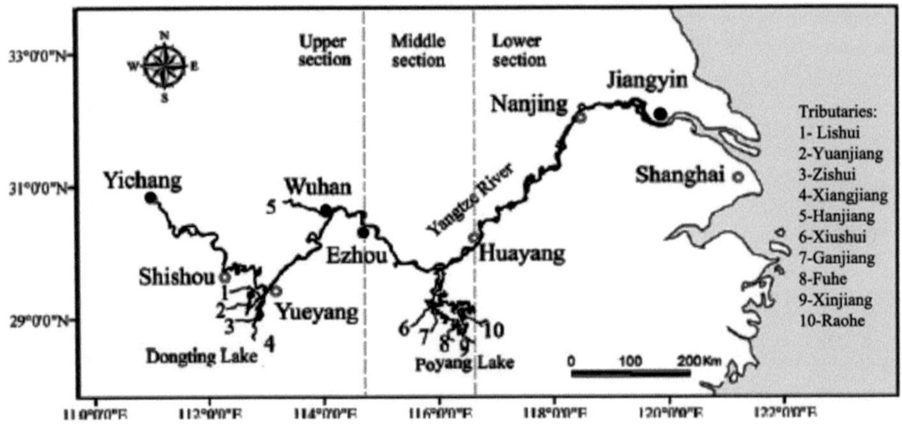

Fig. 2.9 Middle and lower sections of the Yangtze River (Source: Zhao et al. 2008)

with decreasing water temperature and positively correlated with increasing salinity and also with the number of reported neonatal porpoise stranding. Seasonal distribution appears to be linked with reproductive cycles and hydrography; and both diurnal patterns and tidal state seem to affect the abundance (Parsons 1998b).

2.4.4 Protection Status of Yangtze Finless Porpoises

The YFPs are now listed as the Second Order of Protected Animals in China and have been listed as an endangered population in the IUCN Red Data Book since 1996, based on the criteria that fewer than 2,500 mature individuals remained at that time and the population was continuing to decline (Baillie and Groombridge 1996; Wang et al. 2005). Based on surveys conducted between 1984 and 1991, it has been estimated that the YFP population numbered approximately 2,700 in the Yangtze River prior to 1991 (Zhang et al. 1993). However, since that time the population has sharply decreased (Wang et al. 2000; Wei et al. 2002). The latest amount of the YFPs was estimated to be around 2,000 porpoises, based on three range-wide surveys from 1997 to 1999 (Wang 2009; Zhang et al. 2003; Zhao et al. 2008). In the middle and lower regions, between Wuhan and Jiangyin (see Fig. 2.9), porpoise distribution appeared continuous but the abundance had decreased from the level of 1,652 (*surveys of 1984–1991,* Zhang et al. 1993), or 1,481 (*surveys of 1989–1992,* Zhou et al. 1998b), to a current level of around 800 (Zhao et al. 2008).

It is estimated that there are around 1,200 YFPs living in the main river channel (Zhao et al. 2008), around 400 porpoises living in Poyang Lake (Xiao and Zhang 2000), and around 100–150 YFPs living in Dongting Lake (Yang et al. 2000). In sum, there are estimated to be around 1,800 YFPs in the Yangtze River system.

Since the early 1990s, at least half the porpoise population has been lost, implying an annual rate of decline of at least 5 % (assuming the decline was exponential) (Zhao et al. 2008). A population viability analysis suggested that the YFPs will become extinct within 24–94 years if no protective measures are taken (Zhang and Wang 1999).

There is also some data on density and distribution of the YFPs, which indicates the need for assignment of protection status. During the surveys in the spring of 1991 and 1992, the encounter rate in the segment of the Yangtze River between Wuhan and Xinchang declined from 0.14 porpoise/km to 0.07 porpoise/km (Wang et al. 2000). Similarly, the observed density of the YFPs in the subsection from Yichang to Ezhou decreased from 0.11 porpoise/km in 1991 (Zhang et al. 1993) to 0.02 porpoise/km at present, which indicates that the YFPs in this section appear to be at the highest risk of disappearing. If the porpoises in this subsection were to die off, the habitat range of YFPs would shrink by 400 km, or about 24 % (Zhao et al. 2008). The last survey in the section between Wuhan and Yueyang, as the one between Dongting Lake and Poyang Lake (Barrett et al. 2006) indicated that the densities in Dongting Lake and Poyang Lake are relatively higher than that found in the river section, 1.020 individuals/km and 0.085 individual/km, respectively. Additionally, the observations conducted in 2005–2006 indicated that the density of the YFPs in Poyang Lake is considerably higher than that in the adjacent section of the Yangtze River (Wang et al. 2006). However, there are still some threats faced by porpoises in Poyang Lake.

It can be concluded that the porpoise population is decreasing dramatically. The density of the YFPs is reducing in most sections of the Yangtze River system, with the exception of in Poyang Lake. That is to say, the already very limited habit of the YFPs is shrinking. What is causing the decrease of porpoise population and loss of their habitat? As the best habitat for YFPs, what about the situation of the YFPs in Poyang Lake? Are they completely safe? No, the YFPs in Poyang Lake are also facing some existent and potential threats, which are explained in the next section.

2.4.5 *Threats in Poyang Lake*

The threats faced by the YFPs are mainly from the unfettered exploitation of the lake's resources and the severe environmental pressure caused by the human population. The main threats faced by the YFPs refer to the use conflicts above, including overfishing and illegal-fishing, vessel traffic, and underwater noise associated with sand mining, water project development, and water pollution (Reeves et al. 2005; Wang et al. 2006). As for direct threats to the YFPs in Poyang Lake, these are mainly from illegal fishing and overfishing, sand dredging, and water pollution (see Fig. 2.10). In addition, the Poyang Lake Project, now in the planning stages, could potentially produce great threats to the long-term survival of the YFPs.

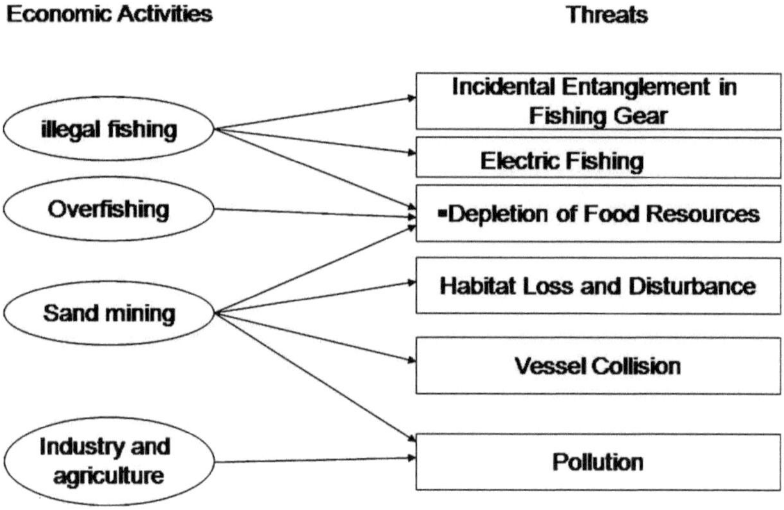

Fig. 2.10 Threats faced by the YFPs in Poyang Lake

2.4.5.1 Overfishing and Illegal Fishing

The YFPs mainly feed on fish and shrimp. At present, the Poyang Lake is overfished by net fishing and trap fishing, which lead to the serious decrease of fishing product. Over-fishing to feed the increasing human population also reduces the availability of YFPs' prey, making it harder for them to get enough food.

By the early 2000s, the illegal use of dynamite fishing and electric fishing were the most important and immediate direct threats to the YFPs' survival and to the other aquatic organisms including the preys of the YFPs (Zhang et al. 2003; Wang et al. 2005). The YFPs were also at risk from a harmful device called a rolling hook, which causes these animals to drown. A rolling hook is a long rope equipped with many large, sharp hooks, which is either dangled in the middle of the water or placed on the bottom of the river to catch fish (Chen et al. 1997). Protection from deliberate killing of the YFPs has been in place for a considerable time, but the banning of harmful fishing appears to be difficult if not impossible, to enforce in the Yangtze River (Zhou et al. 1998a; Reeves and Gales 2006). There are always some people risking illegal fishing for the sake of economic benefits.

Since 2002, spring and winter bans have been imposed for the whole of Poyang Lake, during which periods all fishing boats and tackle are forbidden. The bans have proven effective in restoring fish resources and ecology in the lake over the past decade (Xinhua News Agency 2006). Indirectly, the policy also helps with the protection of the YFPs because more food sources, like fish and shrimp, can be supplied and accidental harm caused by fishing can be avoided in these periods.

2.4.5.2 Sand Mining

Because sand mining has been prohibited in most sections of the Yangtze river, most of the sand dredging vessels have moved into the Poyang Lake. Sand mining has become a mainstay of local economic development in the Poyang lake region in the last few years, and it is an important source of revenue in the region that borders Poyang Lake. There are a few thousand vessels working in the lake for digging and transporting sand (Wang et al. 2006). High density dredging projects have been the principal cause of the death of the local wildlife population, including the YFPs.

Sand mining increases the turbidity of lake water, so that the porpoises cannot see as far they once could and have to rely on their highly-developed sonar systems to avoid obstacles and to search food. Large ships enter and leave the lake at the rate of two a minute and such a high density of shipping causes great noise pollution. The YFPs' sonar systems are badly disturbed by this, as it means that they have difficulty hearing their food and escaping from danger, and also cannot swim freely from one bank to the other (Zhang 2007).

Additionally, the heavy vessel traffic in the outlet channel waters of Poyang Lake might badly affect the migration of the YFPs between Poyang Lake and the Yangtze River (Wang et al. 2006).

Furthermore, sand mining is changing the lake bottom and, therefore, also the habitat of the YFPs. Sand mining machines also suck up large numbers of fish and shrimp, which decreases the amount of prey available for the YFPs. Oil leakage from sand mining machines leads to water pollution, which is harmful to the YFPs. A great number of sand is piled in the inshore area, which is the main habitat range of the YFPs (Yu et al. 2001). This sand piling further shrinks the habitat of the YFPs. It has also been reported that propeller strikes have killed and injured the YFPs (Zhou and Zhang 1991; Chen et al. 1997). In a sentence, sand mining causes threats to the YFPs in many aspects and can be considered as a major threat.

2.4.5.3 Pollution

Pollution may also affect this species. Finless porpoises disappeared from Ise Bay in Japan during a time with high pollution containing DDT isomers and metabolites at high levels and returned when pollution was reduced. There has been some concern about the levels of pollution in the Inland Sea of Japan, where the largest population may be present, at least seasonally (Reyes 1991). From April to June 2004, at least six deaths of YFPs happened in Dongting Lake, which could be related to the presence of heavy metals, organic agricultural chemicals, industrial sewage, and chemical pesticide used to control schistosomiasis (Wang 2009; Wang et al. 2005). The YFPs are a top predator in the food web of the Yangtze River and lakes, and therefore are considered to be the most at risk of the heavy mental poisoning, via accumulation, such as, for example, from mercury. It has been demonstrated that mercury concentration increases positively with the age of the YFPs, and the recorded transfer efficiency of mercury from the YFP mothers to the YFP babies is

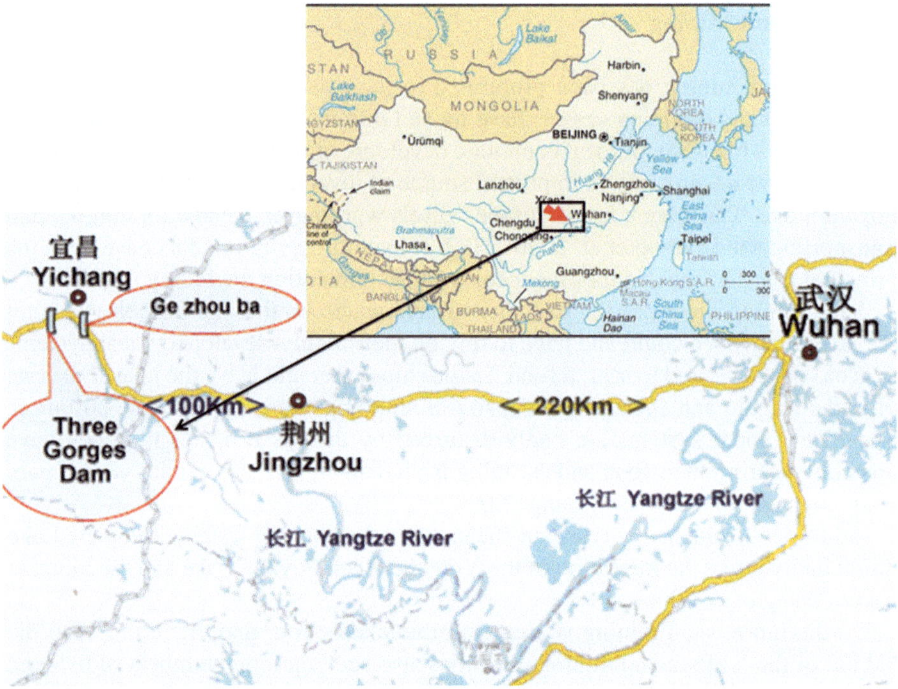

Fig. 2.11 Three Gorges Dam (Source: The Contra Costa Demonstrator, 2002)

remarkable. Enough mercury was accumulated from the aquatic environment to the livers of the YFPs' that it reached a bioaccumulation factor of 4.3×10^5 in the East Dongting Lake (Dong et al. 2006). Although, at present, there are still no detailed statistics about the number of YFPs' deaths caused by pollution in Poyang Lake, it is clear that the control of pollution is a nonnegligible factor for their protection.

2.4.5.4 Water Projects

The increasing rate of water project development, including construction of dams and floodgates in the middle and lower reaches of the Yangtze River, has resulted in blockages between the river and the lakes, and has transformed the YFPs' habitat (Reeves et al. 2005). For the YFPs in the Yangtze River system, the dams have interrupted their movements upstream of the Yangtze River and have eliminated their access to the river's tributaries and connected lakes (Chen and Hua 1989; Liu et al. 2000). The dams also cause a decrease in fish resources (Zhou and Li 1989). The Three Gorges Dam (see Fig. 2.11), which is constructed on the Yangtze River, is the world's largest hydroelectric power generator. The operation of Three Gorges Dam alters the hydrological conditions in the lower reaches of the Yangtze River, which can influence the habitat of the YFPs (Chen et al. 1997).

In addition, the Poyang Lake Project, a new dam building project, is one important development strategy of Jiangxi Province. With this project, first, the habitat of the YFPs in Poyang Lake will be fragmented and reduced (see Fig. 2.4). Secondly, the local hydrological conditions will be changed, which will affect their habitat. Thirdly, the operation of the Poyang Lake Project, upon completion of the dam, will greatly decrease the available population of prey for the YFPs. Fourth, the operation of the hydropower plant will influence the local ecological conditions to a great extent and alter the habitat quality of the YFPs. In one word, the influence of Poyang Lake Project on the YFPs is potentially serious and is an inevitable result of the project proceeding. With this project, Poyang Lake will lose the position of being the best available habitat of the YFPs.

So far, it can be concluded that the YFPs are facing serious threats. The YFPs are the only entirely freshwater—adapted porpoises in the world. As members of the Second Order of National Protected Animals in China, it is necessary to take measures immediately to protect the YFPs.

2.4.6 Protection of Yangtze Finless Porpoises

The endangered status of the YFPs has attracted worldwide attention, and the possible strategies for protecting them from extinction have been discussed extensively at scientific meetings and in the literature over the past 20 years (Zhou et al. 1994; Perrin et al. 1989; Reeves et al. 2000; IWC 2001; Braulik et al. 2006; Samuel et al. 2006). These protection strategies include in situ conservation, ex situ conservation, and intensification of breeding research in captivity. Recently, issues concerning the feasibility and realistic nature of proposed protection strategies have been raised for debate (Wang et al. 2006; Reeves and Gales 2006; Yang et al. 2006).

2.4.6.1 In Situ Conservation

In situ conservation means "on-site conservation." It is the process of protecting an endangered plant or animal species in its natural habitat, either by protecting or cleaning up the habitat itself, or by defending the species from predators. This term refers also to the conservation of genetic resources in natural populations of plant or animal species, such as forest genetic resources in natural populations of tree species. It is increasingly being applied to conservation of agricultural biodiversity by farmers in agro-ecosystems, especially by those using unconventional farming practices (Wikipedia 2010).

Wildlife and livestock conservation are mostly based on in situ conservation. This involves the protection of wildlife habitats. For example, the YFPs can be protected by rehabilitating their habit. Here, the argument would be that the ecological environment of the whole Yangtze River system should be restored in order to protect the YFPs. However, the Yangtze River is called the "Golden Channel" and it plays an important role in the economic development along the Yangtze River. Therefore, it is extremely difficult to ban shipping or to eliminate water pollution completely.

It is more reasonable to aim to conserve a share of Poyang Lake, the best habitat of the YFPs. The best method is to establish a nature reserve there both to protect the habitat itself and to defend porpoises from predators. As already mentioned above, there are about 400 porpoises living in Poyang Lake, which accounts for about 25 % of the total porpoise population. The population size must be sufficient to enable the necessary genetic diversity to survive within the population, so that it has a good chance of continuing to adapt and evolve over time. The area of the nature reserve can be calculated for target species by examining the population density in naturally occurring situations. The reserve must then be protected from intrusion or destruction by human, and against other catastrophes. For this purpose, a number of local residents can be employed as rangers to supervise the habitat protection.

It generally argues that preserving a species in its natural habitat should always be the first priority of species conservation (Yang et al. 1999; Wang et al. 2006). In situ protection is a very important measure because of the two advantages that it brings. One advantage is that it maintains recovering populations in the surroundings within which they have developed their distinctive species specific, properties. Another advantage is that this strategy helps to ensure opportunities for a species to continue the ongoing processes of evolution and adaptation within their environments. However, when in situ conservation is too difficult or impossible, ex situ conservation may be used on some or all of the population.

2.4.6.2 Ex Situ Conservation

Ex situ conservation literally means "off-site conservation." It is the process of protecting an endangered species of plant or animal by removing part of the population from a threatened habitat and placing it in a new location, which may be a wild area or within the care of humans. While ex situ conservation comprises some of the oldest and best known conservation methods, such as colony relocation, it also involves newer, sometimes controversial laboratory methods, like zoos and botanical gardens (Wikipedia 2010).

As for the protection of the YFPs, there are no such zoos or botanical gardens. However, colony relocation is generally rather prevalent and is a popular form of porpoise protection. Some natural reserves have been built at several sites, including Shishou National Reserve, Tian-E-Zhou oxbow seminatural reserve, Honghu-Xinluo national reserve, Zhejiang provincial reserve, and Tongling provincial reserve (see Fig. 2.12). These nature reserves were initially established for the Baiji (River dolphin) conservation. After the Baiji died out, these nature reserves are now being applied for the protection of the YFPs.

At present, Tian-E-Zhou is the most successful example for the ex situ conservation of the YFPs. The Tian-E-Zhou nature reserve was founded in Shishou in 1992. The Tian-E-Zhou oxbow, located near Shishou city in the Hubei province, is a 21 km long and 1–2 km wide old channel of the Yangtze River. Between March 1990 and December 1996, a total of 36 YFPs (17 females, 19 males) were translocated into the reserve and nine YFPs were born there. Of the 45 animals, 11 died, 14

Fig. 2.12 Distribution of nature reserves for the YFPs (Sources: Chris Huh 2006)

escaped into the Yangtze River during the 1996 flood season, and 15 were released (Wang et al. 2000). Under careful management, these animals cannot only survive, but can also reproduce naturally and successfully. By the end of 2007, this group had become an effective population consisting of more than 30 YFPs.

In addition, Tongling Semi-natural Reserve was approved in January 1993. It was established in a small channel between two islands of the Yangtze River near Tongling City in Anhui Province. This channel is approximately 1.5 km long and 200 m wide (Wang et al. 2006). Five animals were introduced into a small channel in 2001, and five calves were born in 2003, 2005, 2006, 2007, and 2009, respectively. There are totally ten porpoises living in the Tongling nature reserve (Wang 2008).

Ex situ conservation, while helpful in humanity's efforts to sustain and protect our environment, is rarely successful enough to save a species from extinction (Li et al. 2002). It has to be used as a last resort or as a supplement to in situ conservation, because it cannot recreate the habitat as a whole: the entire genetic variation of a species, its symbiotic counterparts, or those elements which over time might help a species adapt to its changing surroundings (Wikipedia 2010). Instead, ex situ conservation removes the species from its natural ecological contexts, preserving it under semiisolated conditions whereby natural evolution and adaptation processes are either temporarily halted or altered by introducing the specimen to an unnatural habitat. Furthermore, ex situ conservation techniques are often costly. For example, the establishment of nature reserves comes at great cost and some nature reserves are in bad state due to lack of funds.

Therefore, ex situ conservation should be emphasized without abandoning in situ conservation (Wang et al. 2006). What is more, it is crucial that the ex situ porpoise population is not seen as a substitute for conservation in the wild and that eventual reintroduction or restocking in the Yangtze River is the end objective of the ex situ project. Relocation to Poyang and Dongting Lakes remains the ultimate end objective of the ex situ effort (Wang et al. 2006).

However, because of the conflicts between porpoise conservation and economic development, it is rather difficult to implement large-scale in situ conservation in the YFPs' original habitats, like the whole Yangtze River system. Hence, ex situ conservation is a very important supplement for the in situ protection. In addition to colony relocation, there is another kind of ex situ conservation of the YFPs—artificial breeding.

2.4.6.3 Intensification of Artificial Breeding

A YFP was born on July 5, 2005 at the CAS Institute of Hydrobiology (IHB) in Wuhan, capital of central China's Hubei Province. This is the world's first successful breeding of such an animal in an artificial environment, which will contribute to the ex situ conservation of the endangered species (Wang et al. 2005). The CAS Institute of Hydrobiology started work on breeding the porpoise in captivity in 1995 and has been a worldwide pioneer in the reproduction of freshwater cetaceans in captivity. In total, there are now seven porpoises living in the dolphinarium in the CAS Institute of Hydrobiology.

The above protection methods are classified according to the protection site. However, the implementation of porpoise conservation also depends on some assistant instruments. For example, a monitoring and rescuing network system needs to be developed, for the long-term monitoring in the important river reaches and watersheds, such as the YFP monitoring and rescue network, centered in Jiangxi, which has now taken shape. The first YFP rescue center on China's Yangtze River opened in the eastern Jiangxi Province. This state-level center consists of one central rescue station in the provincial aquatic products scientific institution in the capital city Nanchang and three substations in the counties of Xingzi, Hukou, and Duchang (see Fig. 2.13). The rescue center will cover more than 5,000 km^2 waters of China's biggest freshwater lake Poyang Lake and the Yangtze River's Jiangxi section, measuring 152 km (Xinhua News Agency 2005).

In addition, Poyang Lake should be converted into a National Nature Reserve of the YFPs and more operation funding should be appropriated by the Chinese government and related international organizations. The habitat for the YFPs should be rehabilitated as soon as possible. The YFPs should be classified as a First Category of National Key Protected Wildlife species in China, so that their protection can be strengthened by law.

2.5 Conclusion

In summary, Poyang Lake, as the largest freshwater lake and the last major basin of clean water in China, holds an important position and plays a great part in promoting economic development and safeguarding the quality of local ecological conditions. Poyang Lake contributes to industrial water use, agricultural water use, fishing, and

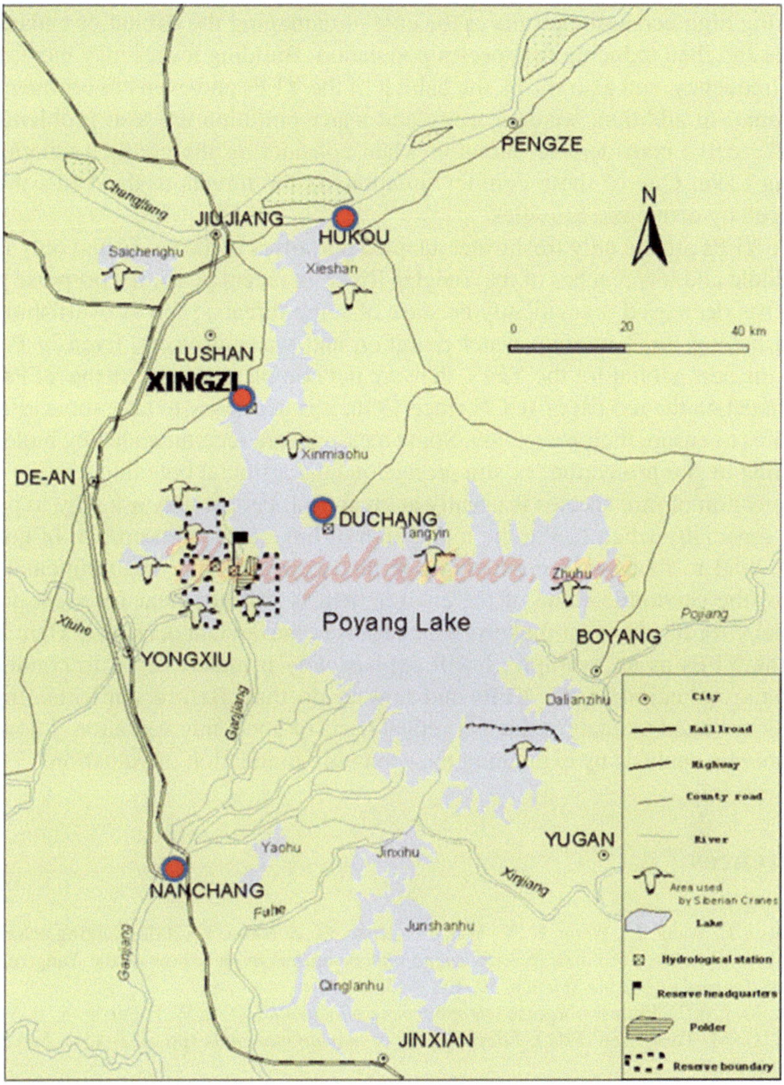

Fig. 2.13 Rescue stations in Jiangxi Province (Source: Huangshan CITS 2008) plus own presentation)

ecotourism, which are classified as economic benefits. There are rich fish resources in Poyang Lake. At the same time, it is also an important habitat for many rare species, like Siberian Crane and the YFPs. Plentiful fish and rare species play an important function in maintaining biodiversity, which can be considered an ecological benefit.

When there is an imbalance between the extraction of economic benefits and the maintenance of ecological benefits, conflicts happens. For example, sand mining can bring high economic profits at the cost of damaging the habitat of endangered species and then reducing the species population. Building levees may increase the flood frequency, and also shrink the habitat of the YFPs and even cut off the migration route. In addition, water scarcity and water pollution are also problems that should receive considerable attention when considering the conflict situations in Poyang Lake. One of these conflict situations is the threats to the YFPs that are produced by economic activities.

The YFPs are the only freshwater adapted porpoises in the world and only live in the middle and low reaches of the Yangtze River. In recent years, the porpoise population has decreased dramatically because of some threats, such as overfishing and illegal fishing, sand mining, water pollution and water projects. Even in Poyang Lake, the best habitat for the YFPs, they are not completely safe. As the YFPs have been listed on the red list of IUCN since 1996, it is necessary to take some effective measures to ensure their protection. Some measures are recommended by biologists, including in situ protection, ex situ protection, and artificial breeding.

Protection of rare species is usually at great cost. People may ask why to protect them, especially when facing the temptation of huge economic profits. In general, people tend to focus on the economic profits while ignoring environmental cost. That is, the economic value of the environment is still not clear for most people, especially for the decision-makers, which leads to the deviant decisions. Here, I will take the YFPs as an example. I will explain why it makes sense to conduct an economic valuation of the YFPs and how to do that. Before implementing the valuation, I need to analyze the theoretical basis of economic valuation. I now turn to economic analysis by explaining the economic foundation of valuation.

References

Akamatsu, T., Wang, D., Wang, K. X., Wang, Z., Zhao, Q., & Naito, Y. (2002). Diving behavior of freshwater finless porpoises (*Neophocaena phocaenoides*) in an oxbow of the Yangtze River. *ICES Journal of Marine Science, 59*, 438–443.

Amano, M. (2002). Finless porpoise (*Neophocoena phocaenoides*). In W. F. Perrin, R. B. Wursig, & J. G. M. Thewissen (Eds.), *Encyclopedia of marine mammals* (pp. 432–435). San Diego: Academic.

Baillie, J., & Groombridge, B. (Eds.). (1996). *1996 IUCN red list of threatened animals*. Gland: IUCN.

Barrett, L., Wang, D., Barlow, J., Pitman, B., Akamatsu, T., Zhao, X., Müller, B., & Pfluger, A. (2006, May 24). *The pilot Yangtze freshwater dolphin expedition report*. San Diego/Wuhan/London/Zurich.

Barzen, J., Engels, M., Burnham, J., Harris, J., & Wu, G. F. (2009). *Potential impacts of a water control structure on the abundance and distribution of wintering waterbirds at Poyang Lake*. Siberian Crane Wetland Project, International Crane Foundation. Web site: http://www.savingcranes.org/poyanglakeproject.html

Bird Life International. (2010). *Species factsheet: Grus leucogeranus*. Retrieved August 6, 2010, from http://www.birdlife.org

Braulik, G. T., Reeves, R. R., Wang, D., Ellis, S., Wells, R. S., & Dudgeon, D. (2006). *Report of the workshop on conservation of the Baiji and Yangtze finless porpoise*. Zurich: Baiji Foundation.

Carwardine, M. (1995). *Whales, dolphins and porpoises*. London: Dorling Kindersley.
Chen, P., & Hua, Y. (1989). *Distribution, population size and protection of Lipotes vexillifer*. Gland: IUCN.
Chen, P., Liu, R., Wang, D., & Zhang, X. (1997). *Biology, rearing and conservation of Baiji*. Beijing: Science Press.
Chen, D., Duan, X., Liu, S., & Shi, W. (2004). Status and management of fishery resources of the Yangtze river. In R. L. Welcomme & T. Petr (Eds.), *Proceedings of the second international symposium on the management of large rivers for fisheries (Vol. I)*. Bangkok: FAO Regional Office for Asia and the Pacific (RAP Publication 2004/17).
China Central TV. (2010). *Poyang Lake's new era of eco-tourism*. Retrieved March 18, 2010, from http://www.china.org.cn/video/2010-03/18/content_19634996.htm
Chris Huh (2006). *Baiji conservation efforts map*. Retrieved from http://en.wikipedia.org/wiki/File:Baiji_conservation_efforts_map.png
Cui, L. J. (2004). Evaluation on functions of Poyang Lake ecosystem. *Chinese Journal of Ecology, 23*, 47–51 (in Chinese).
Culik, B. M. (2004). *Review of small cetaceans: Distribution, behavior, migration and threats* (Regional Sea Reports and Studies No. 177). Retrieved from http://www.unep.org/regional-seas/publications/reports/RSRS/pdfs/rsrs177.pdf
De Leeuw, J., Shankman, D., Wu, G. F., Boer, W. F. D., Burnham, J., He, Q., Yesou, H., & Xiao, J. (2009). Strategic assessment of the magnitude and impacts of sand mining in Poyang Lake, China. *Regional Environmental Change, 10*, 95–102.
Dong, W., Xu, Y., Wang, D., & Hao, Y. (2006). Mercury concentrations in Yangtze finless porpoise (*Neophocaena phocaenoides asiaeorientalis*) from Eastern Dongting Lake, China. *Fresenius Environment Bulletin, 15*, 441–447 (in Chinese).
Ex-situ conservation. Retrieved 2010, from http://en.wikipedia.org/wiki/Ex-situ_conservation
Groot, S. J. D. (1979). An assessment of the potential environmental impact of large-scale sand-dredging for the building of artificial islands in the North Sea. *Ocean Management, 5*, 211–232.
History of evolution—Poyang Lake. (2008). Available from http://en.jxwmw.cn/system/2008/04/30/010040689.shtml
Honkong Business Newspaper. (2009). *Poyang Lake: Who would shoulder the responsibility of safeguarding her magnificences?* Retrieved March 2, 2009, from http://bbs.jjxw.cn/dispbbs.asp?boardid=5&id=56993
Huang, Q. P., & Huang, G. Q. (2007). Question and countermeasure in integrated exploitation and use of water resources of Poyang Lake. *Acta Agriculturae Jiangxi, 19*, 82–85 (in Chinese).
Huangshan CITS. (2008). *Map of poyang Lake*. Retrieved from http://www.huangshantour.com/english/map/poyanglakemap.asp
In-situ conservation. Retrieved 2010, from http://en.wikipedia.org/wiki/In-situ_conservation
International Crane Foundation (1998). *Poyang Lake ecology study*. Retrieved from http://www.savingcranes.org/poyanglakeproject.html
IWC (International Whaling Commission). (2001). Report of the workshop on status and trends of Western North Atlantic right whales. *Journal of Cetacean Resource Management Species, 2*, 61–87.
Jefferson, T. A., & Hung, S. K. (2004). Neophocaena phocaenoides. *Mammalian Species, 746*, 1–12.
Jefferson, T. A., Leatherwood, S., & Webber, M. A. (1993). *FAO species identification guide: Marine mammals of the world*. Rome: UNEP/FAO.
Jiangxi Water Conservancy. (2010). *The poyang lake project*. Retrieved from http://www.jxsl.gov.cn
Kanai, Y., Ueta, M., Germogenov, N., Nagendran, M., Mita, N., & Higuchi, H. (2002). Migration routes and important resting areas of Siberian Cranes (*Grus leucogeranus*) between Northeastern Siberia and China as revealed by satellite tracking. *Biological Conservation, 106*, 339–346.
Kasuya, T. (1999). Finless porpoise – *Neophocaena phocoenoides* (G. Cuvier, 1829). In S. H. Ridgway & S. R. Harrison (Eds.), *Handbook of marine mammals vol. 6: The second book of dolphins and porpoises* (pp. 411–442). London/ New York/Toronto/Sydney/San Francisco: Academic.

Li, T. (2008). *Crisis at Poyang Lake*. Retrieved May 6, 2008, from http://www.chinadialogue.net/homepage/show/single/en/1846-Crisis-at-Poyang-Lake
Li, Q. M., Xu, Z. F., & He, T. H. (2002). Ex situ genetic conservation of endangered *Vatica Guangxiensis* (Dipterocarpaceae) in China. *Biological Conservation, 106*, 151–156.
Liu, Y. (2003). The effect of return grain fields to the lake o the Poyang Lake area. *Jiangxi Science, 21*, 235–238 (in Chinese).
Liu, R., Wang, D., & Zhou, K. Y. (2000). Effects of water development on river cetaceans in China. In R. R. Reeves, B. D. Smith, & T. Kasuya (Eds.), *Biology and conservation of freshwater cetaceans in Asia* (Occasional Paper of the IUCN Species Survival Commission, pp. 40–42). Gland: IUCN.
Min, K. (1998). *Water pollution in China*. Retrieved from http://darwin.bio.uci.edu/-sustain/suscoasts/krismin.html
Parsons, E. C. M. (1998a). The behavior of Hong Kong's resident cetaceans: The Indo-Pacific hump-backed dolphin and the finless porpoise. *Aquatic Mammals, 24*, 91–110.
Parsons, E. C. M. (1998b). Strandings of small cetaceans in Hong Kong territorial waters. *Journal of Marine Biology Association of the United Kingdom, 78*, 1039–1042.
Peng, Z. F. (Ed.). (1999). *Jiangxi province levee atlas*. Jiangxi: Jiangxi Provincial Government Levee Department (in Chinese).
People's Daily. (2000). *China bans illegal sand mining on Yangtze river*. Retrieved May 6, 2008, from http://english.peopledaily.com.cn/english/2000/12/13/eng20001213_57716.html
Perrin, W. F., Brownell, K. Y., & Liu, J. K. (1989). *Biology and conservation of the river dolphins*. World Conservation Union, Gland, Switzerland.
Pilleri, G., & Gihr, M. (1972). Contribution to the knowledge of the cetaceans of Pakistan with particular reference to the Genera *Neomeris, Sousa, Delphinus and Tursiops* and description. In G. Pilleri (Ed.), *Investigations on cetacean*. Berne: der Bund. Iv.
Reeves, R. R., & Gales, N. J. (2006). Realities of Baiji conservation. *Conservation Biology, 20*, 626–628.
Reeves, R. R., Wang, J. Y., & Leatherwood, S. (1997). The finless porpoise, *Neophocaena phocaenoides* (G Cuvier 1829): A summary of current knowledge and recommendations for conservation action. *Asian Marine Biology, 14*, 111–143.
Reeves, R. R., Jefferson, T. A., Kasuya, T., Smith, B. D., Wang, D., Wang, P., Wells, R. S., Wursig, B., & Zhou, K. (2000). Report of the workshop to develop a conservation action plan for the Yangtze river finless porpoise. Ocean Park, Hong Kong, 16–18 September 1997. In R. R. Reeves, B. D. Smith, & T. Kasuya (Eds.), *Biology and conservation of freshwater cetaceans in Asia* (Occasional Paper of the IUCN Species Survival Commission, pp. 67–80).
Reeves R. R., Smith B. D., Wang D., & Zhou K. (2005). Lipotes vexillifer. In IUCN. *2006 IUCN red list of threatened species*, from www.iucnredlist.org
Reyes, J. C. (1991). *The conservation of small cetaceans: A review*. Report prepared for the Secretariat of the Convention on the Conservation of Migratory Species of Wild Animals, UNEP/CMS Secretariat, Bonn.
Rice, D. W. (1998). Marine mammals of the world: Systematics and distribution. In D. Wartzok (Ed.), *Society for marine mammalogy (Special Publication Number 4)*. Lawrence.
Samuel, T., Barrett, A., Braulik, G., & Wang, D. (2006). Implementing the recovery programme for the Yangtze river dolphin. *FFI Oryx, 40*, 258–259.
Secretariat of 11th Living Lakes Conference. (2006). *Sustainable lake management: Balancing agriculture and lake protection*. Paper presented at the 11th Living Lakes Conference, Nanchang. Retrieved from https://www.globalnature.org/bausteine.net/file/showfile.aspx?downdaid=6404&sp=E&domid=1011&fd=2
Shankman, D., & Liang, Q. (2003). Landscape changes and increasing flood frequency in China's *Poyang Lake* region. *The Professional Geographer, 55*, 434–446.
Shankman, D., Keim, B. D., & Song, J. (2006). Flood frequency in China's Poyang Lake region. *International Journal of Climatology, 26*, 1255–1266.
Shirakihara, K., Yoshida, H., Shirakihara, M., & Takemura, A. (1992). A questionnaire survey on the distribution of the finless porpoise, *Neophocaena phocaenoides*, in Japanese waters. *Marine Mammal Science, 8*, 160–164.

Sun, X. S. (2009). *Poyang lake project comprehensively promote the work of water allocation*. Speech on the 3rd River Basin Integrated Management Sub-forum of Yangtze Forum. Water Resources Department of Jiangxi Province, Retrieved April 21, 2009, from http://www.jxsl.gov.cn/article.jsp?columnid=967&articleid=12943

Te Boekhorst, D. G. J., Smits, T. J. M., Yu, X., Li, L., Lei, G., & Zhang. C. (2010). Implementing integrated river basin management in China. *Ecology and Society, 15*(2), 23. Retrieved from http://www.ecologyandsociety.org/vol15/iss2/art23/

The Contra Costa Demonstrator (2002). China Trip Slide Show, retrieved from http://www.ccdemo.info/landabee/ChinaTripSlideShows/512x384/16SS-3GorgesDam/slides/002ChinaMapDams2.html

Travel China Guide. (2011). *Yangtze river facts*. Retrieved from http://www.travelchinaguide.com/river/yangtze/

Vocations To Go. (2011). *Yangtze River*. Retrieved from http://www.vacationstogo.com/images/ports/maps/1043.gif

Wang, D. (2008, July 10). *Problems, practices and perspective of Yangtze cetacean conservation*. Paper presented at the annual meeting of the International Congress for Conservation Biology, Convention Center, Chattanooga, TN. Retrieved June 6, 2010, from http://www.allacademic.com/meta/p243455_index.html

Wang, D. (2009). Population status, threats and conservation of the Yangtze finless porpoise. *Chinese Science Bulletin, 54*, 3473–3484.

Wang, D., Liu, R., Zhang, X., Yang, J., Wei, Z., Zhao, Q., & Wang, X. (2000). Status and conservation of the Yangtze finless porpoise. In R. R. Reeves, B. D. Smith, & T. Kasuya (Eds.), *Biology and conservation of freshwater cetaceans in Asia* (IUCN SSC Occasional Paper No. 23, pp. 81–85). Gland/Cambridge: IUCN.

Wang, D., Hao, Y., Wang, K., Zhao, Q., Chen, D., Wei, Z., & Zhang, X. (2005). The first Yangtze finless porpoise successfully born in captivity. *Environmental Science of Pollution Resource, 5*, 247–250.

Wang, D., Zhang, X., Wang, K., Wei, Z., Würsig, B., Braulik, G. T., & Ellis, S. (2006). Conservation of the Baiji: No simple solution. *Conservation Biology, 20*, 623–625.

Wei, Z., Wang, D., Zhang, X., Zhao, Q., Wang, K., & Kuang, X. (2002). Population size, behavior, movement pattern and protection of Yangtze finless porpoise at Balijiang section of the Yangtze river. *Resources and Environment in the Yangtze Basin, 11*, 427–432.

Wu, G. F. (2008). *Impact of human activities on water level and clarity and underwater light climate of Vallisneria Spiralis L. in Poyang Lake, China*. Doctoral dissertation, International Institute for Geo-information Science & Earth Observation (ITC), Enschede, the Netherland.

Wu, G. F., De Leeuw, J., Skidmore, A. K., Prins, H. H. T., & Liu, Y. (2007). Concurrent monitoring of vessels and water turbidity enhances the strength of evidence in remotely sensed dredging impact assessment. *Water Research, 41*, 3271–3280.

Würtz-Artescienza, M. (2002). *Neophocaena phocaenoides*. Retrieved from http://www.cms.int/reports/small_cetaceans/data/N_phocaenoides/N%20phocaenoides_wurtz.jpg

Xiao, W., & Zhang, X. (2000). A preliminary study on the population size of Yangtze finless porpoise in Poyang Lake. *Jiangxi Chinese Biodiversity, 8*, 106–111.

Xinhua News Agency. (2005). *Jiangxi builds 1st finless porpoise rescue center on Yangtze river*. Retrieved March 26, 2005, from http://news.xinhuanet.com/english/2005-03/26/content_2747790.htm

Xinhua News Agency. (2006). *Fishing Ban starts on Poyang Lake*. Retrieved March 20, 2006, from http://spanish.china.org.cn/english/China/162323.htm

Xinhua News Agency. (2008). *River sand mining banned in largest freshwater lake*. Retrieved April 1, 2008, from http://www.chinadaily.com.cn/china/2008-04/01/content_6583273.htm

Xinhua News Agency. (2009). *China's largest freshwater lake starts three-month fishing ban*. Retrieved March 21, 2009, from http://news.xinhuanet.com/english/2009-03/21/content_11047470.htm

Xinhua News Agency. (2010). *Largest freshwater lake starts fishing ban*. Retrieved March 20, 2010, from http://www.chinadaily.com.cn/china/2010-03/20/content_9618039.htm

Xu, Y. M. (2005). *Research on the ecological water use and water resources optimization allocation in Poyang Lake area*. Doctoral dissertation, Nanchang University (in Chinese).

Xu, D. L., Xiong, M., & Zhang, J. (2001). Analysis of hydrological characteristic of Poyang Lake. *Yangtze River, 32*, 21–27 (in Chinese).

Yang, G., Zhou, K., Xu, X., & Leatherwood, S. (1999). A survey on the incidental catches of small cetaceans in coastal waters of China. *Chinese Journal of Applied Ecology, 10*, 713–716 (in Chinese).

Yang, J., Xiao, W., Kuang, X., Wei, Z., & Liu, R. (2000). Studies on the distribution, population size and the activity of Lipotes vexillifer and Neophocaena phocaenoides in Dongting Lake and Poyang Lake. *Resources and Environment in the Yangtze Basin, 9*, 444–450 (in Chinese).

Yang, G., Bruford, M. W., Wei, F., & Zhou, K. (2006). Conservation options for the Baiji: Time for realism. *Conservation Biology, 20*, 1523–1739.

Yoshida, H. (2002). Population structure of finless porpoises (*Neophocaena phocaenoides*) in Coastal waters of Japan. *Raffles Bulletin of Zoology, 10*, 35–42.

Yu, D., Dong, M., Wang, J., & Zhang, X. (2001). Population status of Yangtze finless porpoise in the Yangtze river section from Hukou to Nanjing. *Acta Theriologica Sinica, 21*, 174–179 (in Chinese).

Zhang, K. (2007). *Poyang Lake: Saving the finless porpoise*. Retrieved May 6, 2008, from http://www.chinadialogue.net/article/show/single/en/839-Poyang-Lake-saving-the-finless-porpoise

Zhang, T. L., & Li, Z. J. (2007). Fish resources and fishery utilization of Lake Poyang. *Journal of Lake Sciences, 19*, 434–444 (in Chinese).

Zhang, X., & Wang, K. (1999). Population viability analysis for the Yangtze finless porpoise. *Acta Ecologica Sinica, 19*, 529–533.

Zhang, X., Liu, R., Zhao, Q., Zhang, G., Wei, Z., Wang, X., & Yang, J. (1993). The population of finless porpoise in the middle and lower reaches of Yangtze river. *Acta Theriol Sinica, 13*, 260–270 (in Chinese).

Zhang, X., Wang, D., Liu, R., Wei, Z., Hua, Y., Wang, Y., Chen, Z., & Wang, L. (2003). The Yangtze river dolphin or Baiji (*Lipotes vexillifer*): Population status and conservation issues in the Yangtze river. *China Aquatic Conservation: Marine and Freshwater Ecosystems, 13*, 51–64 (in Chinese).

Zhao, X. J., Barlow, J., Taylor, B. L., Pitmanc, R. L., Wang, K. X., Wei, Z., Stewart, B. S., Turvey, S. T., Akamatsu, T., Reeves, R. R., & Wang, D. (2008). Abundance and conservation status of the Yangtze finless porpoise in the Yangtze river, China. *Biological Conservation, 141*, 3006–3018.

Zhong, Y., & Chen, S. (2005). Impact of sand dredging on fish in Poyang Lake. *Jiangxi Fishery, 1*, 15–18 (in Chinese).

Zhou, K., & Li, Y. (1989). Status and aspects of the ecology and behavior of the Baiji, *Lipotes vexillifer* in the Lower Yangtze river. In W. F. Perrin, R. L. Brownell, K. Zhou, & J. Liu (Eds.), *Biology and conservation of the river dolphins* (IUCN Species Survival Commission Occasional Paper No. 3, pp. 86–91). Gland: IUCN.

Zhou, K., & Zhang, X. (1991). *Baiji, the Yangtze river dolphin and other endangered animals of China*. Nanjing: Yilin Press. in Chinese.

Zhou, K., Ellis, S., Leatherwood, S., Bruford, M., & Seal, U. (1994). *Baiji (Lipotes vexillifer) population and habitat viability assessment*. Mammalogical Society of China, IUCN SSC Cetacean Specialist Group and IUCN SSC Captive Breeding Specialist Group.

Zhou, K., Sun, J., Gao, A., & Würsig, B. (1998a). Baiji (*Lipotes vexillifer*) in the lower Yangtze river: Movements, numbers, threats and conservation needs. *Aquatic Mammal, 24*, 123–132.

Zhou, K., Yang, G., Gao, A., Sun, J., & Xu, X. (1998b). Population abundance and distribution characteristics of finless porpoise in the river section from Nanjing to Hukou of the Yangtze river. *Journal of Nanjing Normal University (Natural Science), 21*, 91–98 (in Chinese).

Zhu, L., Zhao, Y. W., & Liu, L. M. (2004). Protective utilization and function estimate of wetlands ecosystem in Poyang Lake. *Journal of Soil and Water Conservation, 18*, 196–200 (in Chinese).

Chapter 3
Theoretical Background on Economic Valuation

Abstract In this chapter, the underpinning theory of the economic valuation study will be explained. The goal of this chapter is twofold: first, I want to give an overview of economic valuation as far as it is relevant for the evaluation of the YFPs. Secondly, I will prepare the methodological approach employed in this study.

The structure of the chapter is as follows: I begin by introducing the importance of economic valuation of environmental assets in Sect. 3.1. In Sect. 3.2, I explain why to make an economic valuation of the YFP protection. The reasons are that the protection of YFPs is a public good and is not completely entitled with property rights and that the protection of YFPs produces a number of positive externalities that fail to be internalized. Consequently, market failure happens in porpoise protection. In Sect. 3.3, I describe the total economic value (TEV) of the porpoise protection in the Yangtze River system, including the use values and non-use values. In the following Sect. 3.4, I compare various valuation methods and make the choice of Contingent valuation method (CVM), and in Sect. 3.5, I review the theoretical basis of the CV study. Welfare measures, willingness to pay (WTP) and willingness to accept (WTA) are important theoretical bases for the CV study. Finally in Sect. 3.6, I summarize the challenges faced by the CV and clarify that I will concentrate on addressing the problem of distance decay.

Keywords Economic valuation • Total economic value • Welfare measures • Contingent valuation method • Challenges

3.1 Introduction: Reasons for Economic Valuation

Economic valuation is intended to assess the economic value of the environment. "Valuation can be seen as a method of determining the relative importance of environmental consequences of economic activities. It helps political authorities to make informed decisions about biodiversity conservation" (UNEP 1995). Economic valuation has a potentially wide application, including use in monetized green

Y. Dong, *Contingent Valuation of Yangtze Finless Porpoises in Poyang Lake, China*,
DOI 10.1007/978-94-007-2765-6_3,

national accounts, damage assessments, setting rates for potential environmental taxes, and policy priorities, and for designing projects and programs.

One context in which economic valuation is applied, in particular in the USA and the UK, is cost-benefit analysis. Benefits and costs are both defined in terms of individual's preferences. An individual gain is the benefit of receiving something for which one would like to give up something else in return. To measure how large that benefit is, we measure how much he or she is willing to give up in order to obtain the benefit. Conversely, a person incurs a cost by willingly relinquishing something for which he or she was given something in compensation. To measure how large the cost is, we measure how much he or she is willing to accept in compensation for incurring the cost (Bateman et al. 2002a).

However, benefits and costs do not measure value absolutely. Therefore, one good should be chosen as one unified measurement standard of value for all individuals. The good must be all the people's favorite and should be finely divisible. Conventionally, money can play the role of the good. Money can be put to many uses and can gain more preferences than any other goods. That is, money is a particularly suitable substitute measurement to stand in for goods with greater variability in their preference profiles (Bateman et al. 2002a).

Schulz and Wicke (1986) have summarized five reasons why monetary values for environmental goods should be estimated: (1) Comprehensibility: money demonstrates the social importance of nature conservation programs in an easily understandable way. (2) De-emotionalism: economic valuation (in the form of CBA) demands a complete integration of all benefits and costs, which helps get a better overview of the consequences of projects. De-emotionalism is highly required in eliciting the values of environmental goods in the context of valuating costs and benefits. (3) Measurability. Any investment aims at the biggest profits, as does the investment in the environmental protection. The monetary measurability of environmental damage costs is necessary for making efficient decisions concerning environmental protection. (4) Capability of internalization. Internalization of external effects means that the potential damage on environment should be considered seriously and incorporated into cost (see also Sect. 3.2 below). Economic valuation can provide information about external effects and can thus support environmental policy makers in choosing and designing policy instruments. (5) Amendment of gross domestic product—GDP. Green GDP accounting incorporates not only the economic growth but also environmental damages and health cost. Therefore, monetary valuation of environmental damage is necessary for building up green GDP accounts.

At present, the protection of YFPs in Poyang Lake faces a range of threats, such as overfishing and illegal fishing, sand mining, water pollution, ship traffic, and water projects development (see Chap. 2). Facing the temptation of fruitful profits, both local governments and local residents would prefer to develop economic activities at the cost of losing the YFPs, because the values of the YFPs are ignored. However, there are trade-offs between economic development and the protection of the YFPs. Economic valuation can help make these trade-offs visible. The quantification and detailed explanation of the values of the YFPs is a supporting tool for establishing a balance between economic development and the protection of YFPs so that efficient conservation decisions can be made concerning the management of the YFPs.

Economic valuation should aim at projects rather than species because it provides aids to policy decision. It is useful at levels from political to administrative to the general public. Therefore, in this research my valuation object is the conservation of YFPs in Poyang Lake. However, there is at present no market for the protection for YFPs where people can state their preferences. Market failure happens in the protection of these porpoises, as can be seen in the following section.

3.2 Market Failure on the Protection of Yangtze Finless Porpoises

In general, market failure means that the market does not allocate resources efficiently. Similarly, market failure may mean that the market does not work well on the allocation of porpoise protection because the protection of YFPs is not entitled with well-defined property rights. Instead, porpoise protection is a public good and the protection of YFPs produces lots of external effects that are not internalized in the analysis of the conservation project. In this section, I will illuminate these aspects in more detail. I start with the argument of market failure.

3.2.1 The Theory of Market Failure

Ledyard (1987) notes that "the best way to understand market failure is to first understand market success." The market system is considered successful when a set of competitive markets generate an efficient allocation of resources between and within economies. In economics, efficiency is defined as Pareto optimality—the impossibility of reallocating resources to make one person in the economy better off without making someone else worse off (Hanley et al. 1997).

It is commonly accepted that the Pareto inefficient allocation of goods is to be avoided. When an economic system is Pareto inefficient, there is theoretical potential for a Pareto improvement. Pareto improvement means that a certain change in the allocation of goods may result in some individuals being made "better off" without making any individual worse off. Here "better off" is often interpreted as "put in a preferred position." Conversely, when no further Pareto improvement can be made with the market allocation of some scare resource, the outcome of the market is Pareto efficient.

On the basis of the concept of market success, a market failure exists when the allocation of goods or services by the market is not efficient (Sidgwick 1885, 1901; Bator 1958). That is, Pareto efficiency is not satisfied in the situation of market failure. In the situation of market failure, it is possible to make Pareto improvement through reallocation.

The theorem of welfare economics summarizes the major benefits of markets for social welfare, of which the first fundamental theorem is of most concern for market failure. Four conditions for market success are listed in the first theorem, including (1) a set of complete markets with well-defined property rights; (2) competitive transaction of maximizing benefits and minimizing costs; (3) transparent market prices; and (4) zero transaction costs (Bator 1958). When these four conditions are satisfied, the allocation of resources will be Pareto efficient. A market failure occurs when one of the four conditions is not met, and the allocation of resources is consequently inefficient.

A key requirement to avoid a market failure is the existence of a set of complete-enough markets that cover each and every possible transaction or contingency so that resources can move to their highest valued use (Hanley et al. 1997).

No market, where suppliers and consumers can exchange it, exists for the protection of the YFPs. Obviously, the key requirement for Pareto optimality is not satisfied in the protection of YFPs. Consequently, no competitive transaction occurs in porpoise protection. At present, the Chinese government offers the porpoise protection and can be regarded as the supplier. However, nobody is interested in buying the porpoise protection. There is no data about the market price and transaction cost of the porpoise protection, either. In summary, none of the conditions for market success are met for the protection of YFPs. Explicitly, market failure is present in the area of porpoise protection.

According to the mainstream economic analysis, market failure can occur for one of three reasons, such as monopoly, externalities, and the attributes of public goods (DeMartino 2000). The protection of YFPs is a public good. More fundamentally, the underlying cause of market failure is that the porpoise protection is not entitled with well-defined property rights. At the same time, the porpoise protection can have externalities, which can be attributed to its features as a public good. The following sections explain the reasons for this particular instance of market failure in more detail. Special attention will be paid to the public good character of the porpoise protection and associated externalities.

3.2.2 Public Goods

The conception of public good is determined by the choice of criteria. Economists usually use several criteria for defining public goods. The first criterion is non-rivalry in consumption. Non-rival goods can be simultaneously consumed by many people, like clean air. One person's consumption does not reduce the quantity or quality of that good consumed by others (Samuelson 1954). That is, when one individual utilizes one public good to satisfy his or her own wants, this does not reduce the utility that others can gain from the good. For example, if one person goes to Poyang Lake for the sightseeing of YFPs, this neither prevents anyone else from enjoying the porpoise sightseeing nor lessens the quality of the sightseeing by other beholders.

Table 3.1 Types of goods

	Rivalry	Non-rivalry
Excludability	Private goods Examples: food, clothing, cars	Club goods Examples: satellite television
Non-excludability	Common goods Examples: fish, hunting game, water	Pure public goods Examples: national defense, air

Source: Samuelson (1954, 1955)

A second criterion is non-excludability. A non-excludable good can be enjoyed by anyone, whether he pays for it or not. It means that it is impossible for the supplier of the good to exclude anyone from using perfectly non-excludable goods. People have equal access to the good, which is not affected by the values that they place on the good (Malkin and Wildavsky 1991). For instance, everybody can have equal access to the porpoise sightseeing whether he contributes to the conservation of YFPs or not. Nobody can be excluded from seeing them because it is impossible to raise a fence around Poyang Lake.

These two criteria are most commonly used by economists. Moreover, some economists add even more criteria. For instance, Stiegler gives a third criterion of impossibility of rejection. Impossibility of rejection means that an individual cannot decline consumption of a public good even though he may wish to; a pacifist is protected by nuclear weapons in spite of his moral objections to them (Stiegler 1976). Again, whether the air is polluted or not, people must breathe. The protection of YFPs holds important ecological function, such as maintaining the ecosystem stability and fostering unanticipated promotion of biodiversity. Whether people care about it or not, they are benefiting from the porpoise protection (in this case: biodiversity) and they cannot reject it either.

A further criterion is that for a good to be public, everyone must consume the same amount of it; each person will receive the same amount of national security from the army as every other resident of a country (Samuelson and Nordhaus 1989; Varian 1987).

Most economists concentrate on the two criteria of non-rival and non-excludable and classify the goods according to Table 3.1. In addition to private goods and public goods, Table 3.1 points to common goods and club goods.

Common goods raise similar issues to public goods. They are characterized by rivalry in consumption and non-excludability. The mirror to the public goods problem for this case is called "tragedy for the commons." The "tragedy of the commons" was put forward by Garrett Hardin in 1968. It refers to a dilemma in which multiple individuals acting independently in their own self-interest can ultimately destroy a shared limited resource even when it is clear that it is not in anyone's long-term interest for this to happen (Hardin 1968). For example, a group of herders share a parcel of meadow. Every herder puts as many cows as possible onto the meadow for the sake of one's self-interest. As a result, the common meadow is damaged.

Table 3.2 Classification of goods produced by the protection of Yangtze finless porpoises

	Rivalry	Non-rivalry
Excludability	Meat...	Scientific research
	Pharmaceutical utility...	Education
Non-excludability		Recreation
		Ecological function
		Option value
		Bequest value for future generations
		Existence value

Source: Own presentation

Club goods are excludable but non-rival, which means that while certain people can be excluded from the consumption of a good, one person's consumption of it does not diminish another person's consumption. Club goods are sometimes classified as a subtype of public goods (Buchanan 1965). When the goods are non-rival and non-excludable, they are classified as pure public goods.

The protection of YFPs produces several kinds of goods or products that hold different attributes in terms of rivalry and excludability. Table 3.2 shows that different goods and services associated with the porpoise protection hold different characteristics.

The economic values in Table 3.2 will be explained in the subsequent section. The meat of porpoises is edible. The grease of porpoises can be used for curing burned skin. The blubber can be made into nutritional goods. For these goods, one individual's consumption definitely reduces and excludes any other's consumption of the goods. Now, however, the YFPs are a Second Order of National Protected Animals and it is not allowed to exploit porpoise population for these goods. That is to say, these goods are not gained from the porpoise protection in the current situation at present.

The protection of YFPs also serves for scientific research and education. Bionic research, behavioral research, genetic research and protection research are all of great practical importance. If people want to do some research on the porpoise protection in Poyang Lake, especially when they want to catch some porpoises for research, they have to ask for the permission of Jiangxi Poyang Lake Administration Authority. The same requirement applies for the educational function. Hence, the benefits of scientific research and educational function are non-rival but excludable and can be regarded as club goods.

The other benefits, such as recreation and ecological function are both non-rival and non-excludable, and therefore can be taken as pure public goods. Putting the unexploited goods produced by the porpoise protection on one side, the other products from the porpoise protection are definitely not private goods. Therefore, the protection of YFPs can be taken as a bunch of public goods. Unavoidably, the tragedy of commons happens with the protection of YFPs. The original reason is that the porpoise protection is not entitled with well-defined property rights so that people lack the responsibility and motivation to contribute to it.

Another perspective on public goods, with rather a quite different view, is the perspective of property rights. As this perspective addresses the responsibilities for resource use, it will be addressed in the following section.

3.2.3 Property Rights

Property is not an object but rather a social relation that defines the property holder with respect to something of value (the benefit stream) against all others. Society approves the uses selected by the holder of the property right with governmental administered force and with social ostracism (Alchian 1965). Property is a triadic social relation involving benefit stream, rights holders, and duty bearers (Hallowell 1943; Bromley 1991). A property right is the capacity to call upon the collective to stand behind one's claim to a benefit stream. A right only has effect when there is some authority system that agrees to defend the right holder's interest in a particular outcome (Bromley 1991). A well-defined property right system represents a set of entitlements that define the owner's privileges and obligations for use of a resource or asset and have the following general characteristics: it is comprehensively assigned, exclusive, transferable, and secure (Hanley et al. 1997).

Schlager and Ostrom (1992) identify five important rights that are bundled together to define one property regime, including access, withdrawal, management, exclusion, and alienation.

- Access is the right to enter a defined physical property. If a group of people holds rights of access, they have the authority to enter a resource. For example, regarding the porpoise protection, the access-holder can have authority to enter the resource and enjoy the benefits brought by the porpoise protection under the precondition of not subtracting others' benefits.
- Withdrawal is the right to obtain the products of the property. The people who hold the right of withdrawal on the porpoise protection could have the right to obtain the products of the porpoise protection, such as the pharmaceutical products.
- Management is the right to regulate use and improvements. Individuals who hold rights of management have the authority to determine how, when, and where harvesting from a resource may occur, and whether and how the structure of a resource may be changed (see, e.g., Davis 1984). For instance, the people, who implement the porpoise protection and limit some types of harvesting activities, are exercising rights of management for their resource.
- Exclusion is the right to determine who has access and who can be excluded from using the property. Individuals who hold rights of exclusion have the authority to define the qualifications that individuals must meet in order to access a resource. For instance, the people who limit access to the porpoise protection to certain groups of individuals are exercising a right of exclusion.
- Alienation is the right to sell or lease. Exercising a right of alienation means that an individual sells or leases the rights of management, exclusion, or both.

Table 3.3 Bundles of rights associated with YFPs

	Owner	Proprietor	Claimant	Authorized user
Access	×	×	×	×
Withdrawal	×	×	×	×
Management	×	×	×	
Exclusion	×	×		
Alienation	×			

Source: Schlager and Ostrom (1992)

The people who hold the power of alienation for the protection of YFPs can sell or lease the rights of management and/or exclusion from the porpoise protection.

These five property rights are independent of one another, but in relation to the protection of YFPs, are frequently held in the cumulative manner arrayed in Table 3.3. It is possible to have access rights without withdrawal rights, to have withdrawal rights without management rights, to have management rights without exclusion rights, and to have exclusion rights without the rights of alienation. In other words, individuals or collectivities may, and frequently do, hold certain property regimes that do not include the full set of rights defined above. On the other hand, to hold some of these rights implies the possession of others. For example, the exercise of withdrawal rights is not meaningful without the right of access (Schlager and Ostrom 1992). Following this characteristics of five property rights, four different types of bundles of right can be distinguished:

- Authorized users: the individuals who hold the operational rights of access and withdrawal. At present, everybody is free to access to the porpoise protection in Poyang Lake, China. At the same time, they are free to obtain the benefits brought about by the porpoise protection, such as the sightseeing of porpoises. Authorized users lack the authority to devise their own harvesting rules or to exclude others from gaining access to porpoise protection. For example, the authorized users are not allowed to catch the porpoises for eating. No tourist can exclude other tourists from sightseeing of porpoises.
- Claimants are defined as the individuals who possess the same rights as authorized users plus the right of management. With the right of management, claimants have the authority to devise rights of withdrawal. They cannot, however, specify who may or may not have access to resources, nor can they alienate their right of management. The Administration Authority of Poyang Lake of Jiangxi is authorized with the right of management of the porpoise protection, and can regulate the use and improvements on the porpoise protection. For example, illegal catching of YFPs is forbidden and punished by the Administration Authority of Poyang Lake of Jiangxi except for authorized research institutes, which can catch some porpoises for research. The Administration Authority can also carry out some measures to strengthen the porpoise protection, such as the control of threats to the porpoise protection caused by illegal fishing but cannot

transfer the right of management of the porpoise protection. They cannot do anything with the situation that everyone has access to the porpoise protection.

- Proprietors are defined as individuals who possess rights to participate in management and exclusion. Proprietors authorize who may access resources and how resources may be utilized; however, they do not have the right to alienate either of these rights. Scholars who have recently undertaken theoretical and empirical research on "common-property regimes" focus primarily on those regimes organized by proprietors (National Research Council 1986; Berkes 1989; McCay and Acheson 1987; Ostrom 1990). To use the same term for regimes composed of proprietors, who possess four bundles of property rights, and regimes composed of individuals who possess no property rights, clearly confounds the capacity to communicate about important scientific and policy issues (Schlager and Ostrom 1992). As for the porpoise protection, all the people are free to access it. In other words, nobody has the right of exclusion. But the Ministry of Agriculture has the right to decide how the porpoise protection can be utilized. I can regard the Ministry of Agriculture as the proprietor of the porpoise protection.
- If in addition to rights of management and exclusion, individuals also hold the right of alienation, that is, they can sell or lease their rights of management and exclusion, they are defined as "owners." In China, according to the third item of the first chapter of Law of the People's Republic of China on the Wildlife Protection, all the wildlife resources are owned by the state, including the YFPs. That is, only the state has the right to sell or lease the rights. But because China takes the socialist public ownership as the mainstay, the ownership of the porpoise protection is not transferrable.

In summary, all the people are authorized users and nobody can be excluded. That is, the right of exclusion does not exist. What is more, the porpoise protection is owned by the state and the right of management is not transferrable. Therefore, it can be concluded that, no well-defined property rights system is entitled to the protection of YFPs. Without well-defined property rights, the protection of YFPs produces externalities that are not internalized, which is a great block for the efficient protection.

3.2.4 Externality

In addition to public goods and property right considerations, the concept of externality is widely used to analyze negative environmental effects that are not taken into account by resource uses.

In economics, an externality is an impact on any party who is not directly involved via price mechanisms in a given economic decision. Externalities occur when one person's actions affect another person's well-being and the relevant costs and benefits are not reflected in market prices. In other words, the producers and consumers in a market either do not bear all of the costs or do not reap all of the benefits of the economic activity (Buchanan and Stubblebine 1962).

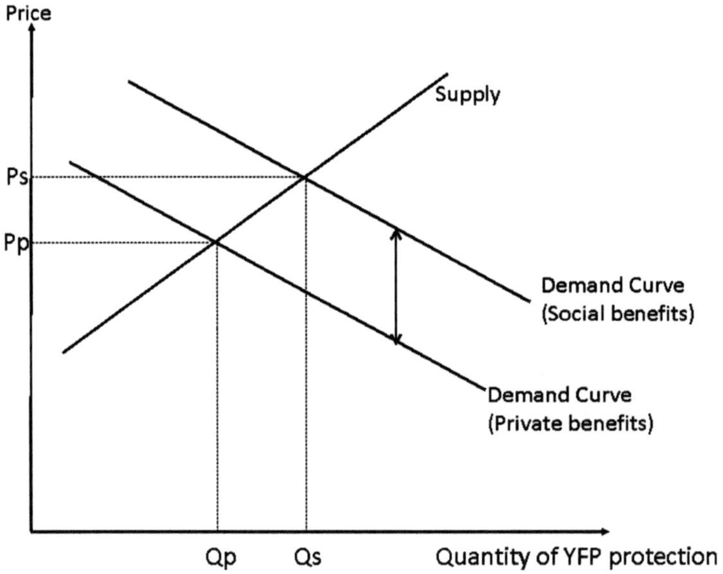

Fig. 3.1 External benefits produced by the porpoise protection (Source: McConnell and Brue 2005)

In general, there are two kinds of externality: positive and negative. A positive externality can be defined as an unpaid benefit enjoyed by others in society that is generated as a by-product of production and consumption. Positive externalities are also known as external benefits (Samuelson and Nordhaus 2005). For example, YFPs are unique rare species in the world, and, therefore, people all over the world can benefit from their protection of their ecological function. But they do not have to pay for that, especially the people outside the Poyang Lake region or even abroad. Figure 3.1 shows the positive externalities produced by the protection of YFPs.

In Fig. 3.1, the supply refers to the protection of YFPs. The effectiveness is quantified by the amount of porpoises. People are presumed to have preferences for the protection of YFPs. People' demand is satisfied by the existence of lots of YFPs. We assume that the stakeholders of the protection project make efforts on the conservation and then can enjoy the benefits from the conservation of the YFPs, including the use values, option values, and non-use values, which are regarded as private benefits. These uses will be explained in the subsequent sections. At the same time, the option values and non-use values are non-excludable and open access to everyone, though these values are not conceived by people precisely. Meanwhile, the porpoise protection can improve the lake ecology and promote the conservation of rare species, like the water fowls. People can enjoy the non-use values and accessory benefits gained from porpoise protection no matter whether they pay for or not. As a matter of fact, the total benefits, namely social benefits, produced by the porpoise protection equal the total values enjoyed by the stakeholders and the non-use values enjoyed by non-stakeholders. External benefits are the non-use values enjoyed by outsiders. By comparing the equilibrium points, I can identify when external benefits are taken into account in the market transaction; the equilibrium point comes to a higher price and a bigger demand, which cannot be realized in the real market. Positive externalities lead to oversupply of the porpoise protection at a higher price. Therefore, market cannot realize the efficient allocation of the porpoise protection in the presence of external costs. Market failure happens in the porpoise protection when external benefits exist.

A negative externality can be defined as an uncompensated harm to others in society that is generated as a by-product of production and exchange. Negative externalities are also known as external cost (Samuelson and Nordhaus 2005). For the moment, the problem of negative externalities needs urgently to be taken into account in order to design effective measures for the conservation of YFPs. Poyang Lake is the last excellent habitat for the YFPs. The successful conservation of YFPs in Poyang Lake is crucial for the sustainability of YFPs in the world. According to biologists' suggestions, building up a nature reserve in Poyang Lake would be the most efficient protection method. But this entails a great cost. In addition to the costs for establishment, management, and operation of a nature reserve, there are also opportunity costs that are often ignored, such as economic loss caused by limiting fishing and banning of sand mining (opportunity costs). The social cost of the porpoise protection equals the private cost to fisher folk and sand miners, plus the cost of forbidding sand mining and limiting the fishery. Then the equilibrium point comes to a higher price and less supply. The price in the marketplace creates an incentive for consumers to over-demand the porpoise protection at a relatively lower price than the optimal price in question. The market is not optimally allocating the porpoise protection when there are unresolved negative externalities. Figure 3.2 shows the external costs attached with the protection of YFPs.

There are a number of potential means for improving overall social utility when externalities are involved. The market-driven approach for correcting externalities is to "*internalize*" third party costs and benefits, for example, by requiring a polluter to repair any damage caused. Another way of internalizing external costs is by Pigouvian taxation. In many cases, however, internalizing external costs or benefits is not feasible, especially when the true monetary values cannot be determined. Therefore, we need to quantify the monetary value of the protection of YFPs by analyzing their total economic value.

3.3 Total Economic Value

3.3.1 Concept of Total Economic Value (TEV)

Economists have generally settled their debates concerning a taxonomy of environmental value, the components of which add up to total economic value (TEV). The key distinction made is between use values and the remainder called non-use values.

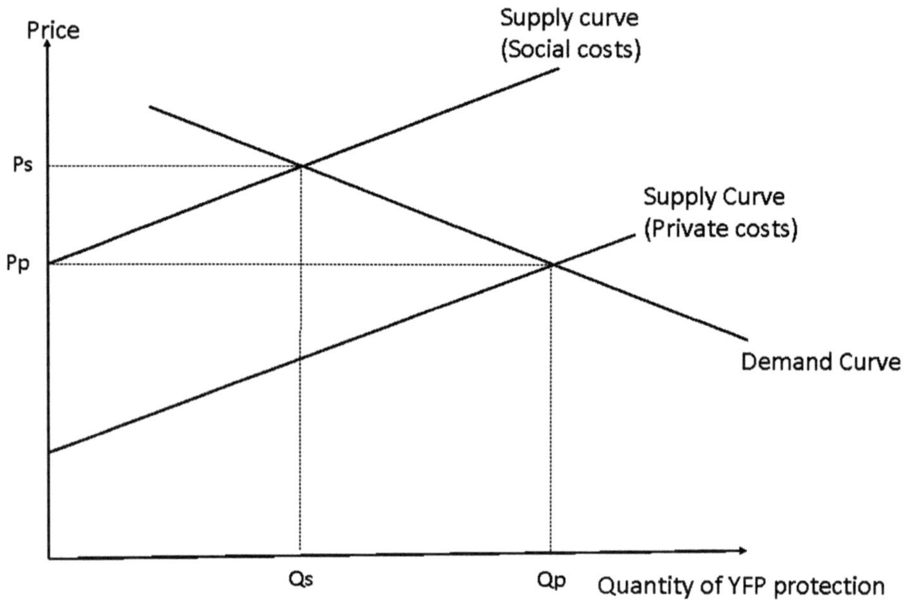

Fig. 3.2 External costs caused by the protection of Yangtze finless porpoises (Source: McConnell and Brue 2005)

Put simply, non-use values cover situations in which individuals who do not use at present, or intend to use in the future, an environmental asset or attribute would perceive it as a "loss" if this asset or attribute was to disappear. They may just wish to see various environmental entities conserved "in their own right" (termed existence value); or conservation may be supported on the basis of retaining options and opportunities for one's children, grandchildren, and future generations (termed bequest value) (Turner 1999).

1. *Use values* relate to actual use of the good in question, planned use or possible use. For example, a visit to see the YFPs is the actual use of the YFPs, and a visit planned in the next 5 or 10 years is planned use. Possible use means that people may want to maintain the YFPs in existence in order to preserve the option of visiting porpoises in the future. Option value thus becomes a form of use value.
2. The notion of *option value* introduced by Weisbrod (1964) is defined as the price that individuals are willing to pay for conservation of an element in view of its possible use in the future (Plottu and Plottu 2007). The users can themselves be the concerned parties here or the option may be concerning the ability of other parties. When the individual is concerned that the good in question should be available to others in the current generation, the value is called *altruistic value*, which is also subsumed under option value. For example, people may like to pay for the protection of YFPs so that they can preserve the potential use of the YFPs in the future. Even though they do not have any possible use of the YFPs

themselves, they think that others could possibly utilize them in the future. In the case, they would like to contribute to the porpoise protection.

3. *Non-use values* refer to WTP to maintain some good in existence even though there is no actual, planned, or possible use. These values have been elucidated by Krutilla (1967), as being those where an individual gets benefit from knowing that an environmental element will be preserved, independently of any personal present or future use. The non-use values show the satisfaction that individuals derive from the knowledge of the existence of environmental assets per se (*existence value*), for the pleasure of others (*altruistic value*) or for future generations (*bequest value*) (Plottu and Plottu 2007).

   Existence value refers to the WTP to keep a good in existence in a context where the individual expressing the value has no actual or planned use for himself or herself or for anyone else. It is the value of the benefits derived from the asset's existence alone. That is, it refers to the intrinsic value of some assets. Normally, natural/environmental motivations here could vary and might include having a feeling of concern for the asset itself or a stewardship motive whereby the appraiser feels some responsibility for the asset. Existence value is separate from the value accruing from any use of the asset (Canada Privy Council Office 2001; Stevens et al. 1991). For instance, people would contribute here toward the protection of the porpoises only because they desire that the YFPs can continue to live in the world.

   A *bequest value* is similar to altruistic value but the concern is that the next and future generations, children and children's children, should have the option to make use of the good in question. It is the value of satisfaction derived from preserving a natural environment for future generations (Greenley et al. 1981). For example, people are willing to pay for the protection of YFPs because they hope their children or others', in successive generations, could have the chance to see the YFPs.

3.3.2 *The Total Economic Value (TEV) of Yangtze Finless Porpoises*

The TEV of the YFPs consists of use values, option values, and non-use values. The TEV of the YFPs is potentially great, as is illustrated below:

3.3.2.1 Use Values

Direct Use Values

The YFPs are of very high direct *economic value*. Its meat is edible. That is, it can be consumed by human or exchanged at the local, household level and even internationally. In 2008, there was a news report concerning the illegal transaction of YFPs in Nanjing. The price was 160,000 RMB per porpoise and 800 RMB per kg on an average, equaling about 90 Euro per kg (Hai 2009). Blubber, oil, and other commodities, such as skin, can be sold in national and international markets at high prices. For example, the skin can be made into leather, by tanning. Blubber and oil are of high nutritional value. The YFPs can also be introduced into aquariums for sightseeing. Because, on the whole, these are marketable commodities, determining appropriate economic values should be relatively straightforward.

In addition, the YFPs have great potential for *pharmaceutical* applications (Simpson et al. 1996). The YFPs contain many biologically active compounds of considerable potential value. Some uses have been known for some time. For example, the grease is a specific medicine for curing burns. More pharmaceutical applications are possible and need to be exploited, which still requires lengthy research at high costs. The YFPs are also a source of genetic material, though once again, predicting potential economic values is difficult in this case.

There is no doubt that *tourism* yields great direct financial benefit. It is more practical because finless-porpoise oriented tourism can be promoted as a "non-consumptive" or "low-consumptive" use of cetaceans that promises monetary rewards to people without requiring that animals be killed or removed from their natural environment. One observatory ferry for the YFPs has been built in Tian-E-Zhou nature reserve in June 2009. If it turns out to be a big success, such observatory ferries could be built up in other parts of the Yangtze River, such as Poyang Lake. Moreover, domestication of the YFPs in order to give porpoise performance is also in consideration. Of course, it is still important to make sure that finless porpoise-watching is conducted in a manner that is respectful of the animals, local human communities, and fellow tourists, guidelines and codes of conduct are increasingly available, and should be adopted and promoted by the tourism industry and by government agencies (e.g., IWC 2002). Potentially, ecotourism can bring about lots of economic benefits.

Like tourism, *scientific research* can generate significant potential quantifiable revenues for local economies. Many economic benefits may ensue from research findings, but immediate financial benefits can be determined through the analysis of annual expenditures or budgets of research projects on the YFPs. The YFPs navigate with a sonar system, so bionic research on the sonar system is important in military science. The research on the blubber and oil is important in nutritional science and burns-curing (Hai 2009). The YFPs are the world's only freshwater-adapted population of the porpoise species, and the genetic research potentially has great economic value, which is currently unknown. Biomedical research can increase the range of commercial products available and help in combating disease.

As for *education*, probably the greatest value arises in the form of *social benefits*. These include the first-hand experience and knowledge gained about the YFPs, together with the increased sense of environmental awareness acquired. By watching the YFPs in Poyang Lake, people's preference for them and some other rare species may be motivated. To determine the extent of this benefit, an adapted form of CVM survey could be carried out, assessing people's opinions before and after visiting the YFPs.

Indirect Use Values

First, the YFPs are the world's only freshwater-adapted population of porpoises. The uniqueness of the YFPs assigns them a special status among all kinds of porpoises. Precise valuation of such benefits is difficult. According to some biologists' opinion, the value of such rare species is priceless. However, I can use WTP to assess a rough value.

Secondly, the YFPs are mammals, the same as human beings. They lie on the top of food chain of the *river ecosystem*. They interact with other elements of the ecosystem in various ways. Their existence plays a key role in maintaining the stability of the whole ecosystem. If the YFPs died out, the biota of lower level elements in the food chain would first increase and then decrease to reach a new balance. It is also possible that other species would die out as a result. The biodiversity of the ecosystem would be influenced greatly. Loss of benefits could potentially be great.

Thirdly, the YFPs are an indicator of *river ecosystem health*. If the river ecosystem cannot support the living of the YFPs, the ecosystem probably cannot support the human being either. Because of the biological concentration effect, people should be at the same level of exposure as the YFPs or even a higher level, within the ecosystem. Correspondingly, the pollutant concentration in human body will be the same as that in the body of YFP, or even higher. But human beings do not necessarily hold stronger immunity system than the YFPs. Namely, if Poyang Lake is not fit for habitation by YFPs, it is possible that it is not suitable for human beings either.

Fourthly, the fish consumption of the YFPs has great *effects on the fisheries*. This impact is closely related with direct use values, but it goes one step future. To better understand the possible effects of this consumption, the knowledge should be improved in a number of areas, such as variations in abundance, distribution, diet, energy requirements, and prey abundance, the way in which the YFPs select their prey and the extent of consumption of fish species by other predators in the ecosystem. The effect as such is sure, but it is a tough task to go into detail about the effect of porpoises on fishery. As a result, it is more difficult to calculate such benefits. But I can choose a simple way to predict the rough value.

3.3.2.2 Option Value

Option value is the benefit received by retaining the option of using a resource in the future by protecting or preserving it today. It is not the expected value from future use, but what people are willing to pay now to maintain their option to use the resource at an undetermined future time. In effect, it is equivalent to an insurance premium guaranteeing the supply of something in the future, which may otherwise become unavailable. For example, some people have never seen a YFP but would like to pay for the protection of YFPs, because they want to keep the option that they may go for sightseeing of YFPs in the future. They are not sure that they will go there some day; but hope that they can still have the chance to see the YFPs if they decide that they want to go to Poyang Lake for the sightseeing of YFPs. To guarantee the existence of this option, they would like to pay for the porpoise protection. Or else, even if they do not plan to visit them on their own, they think other people who are interested in the YFPs should have the option to go to Poyang Lake to visit them in the future. This is similar with other uses, such as pharmaceutical use, genetic resources, education, research, and so on. In addition, the sustainability of YFPs is closely related to the sustainable development of the entire Poyang Lake ecosystem. So the present protection is urgent and necessary, which is essential for the exploitation of their future use. People are willing to pay for the porpoise protection as a guaranteed for all these possible future uses, which should also be ascribed as option value.

Since the YFPs are unique and irreplaceable rare species, it can be assumed that they have great option value. The quantity, uniqueness, and decrease rate will affect option value.

3.3.2.3 Non-use Value

The non-use value of the YFPs mainly includes *bequest value* and *existence value*. The existence value of the YFPs can be defined as the utility that people gain simply from the fact that the YFPs exist and will continue to exist in the future. Part of this satisfaction comes from the fact that people can be sure that future generations can also enjoy the existence of the YFPs, which should be referred to the bequest value.

After having explained the different economic values that are attributed to the YFPs, I now turn to the relationship between values and valuation technique. The technique used in the study reported upon here will then be explained in the subsequent section.

3.3.3 Relationship Between Values and Valuation Technique

The various kinds of values discussed above are illustrated in Fig. 3.3. Starting from the left-hand side, values are highly tangible, and become less and less tangible toward the right-hand side. This determines which valuation technique is most appropriate in each case (OECD 1995):

Direct use values can be estimated using any of the available valuation techniques, such as market valuation of physical effects (MVPE), hedonic price method, travel cost method, contingent valuation, and discrete choice methods. Compared to other types of values, direct use values are the most amenable to market valuation, although the existence of consumer surplus means that the use of prices alone will normally underestimate benefits. Indirect use values can also be estimated using market-based methods, as well as asking people about their willingness-to-pay (WTP). Option, bequest, and existence values can be effectively be discovered only from surveys of people's preferences, expressed in their WTP.

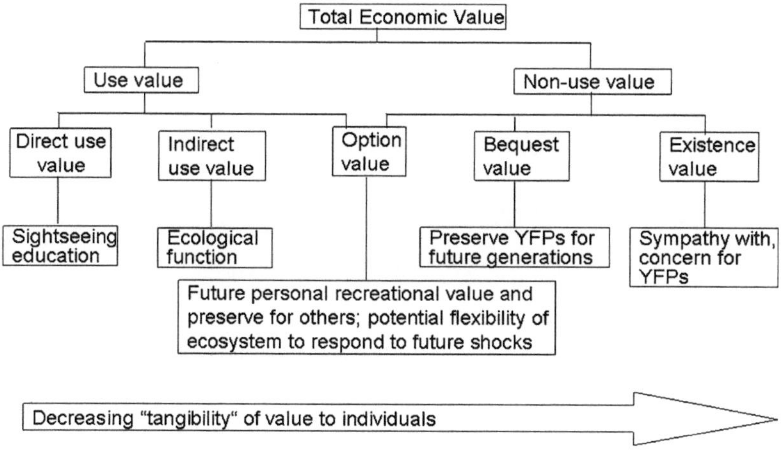

Fig. 3.3 Categories of economic values attributed to environmental assets (Source: Munasinghe 1993)

As can be seen, which values under consideration is very important for choosing a proper valuation method for the case study. In general, any valued object has both use values and non-use values. The difference is which kind of use values take up a greater part of the TEV. This distinction determines what kind of valuation method should be used. In the following section, I will go into detail on various valuation methods and select the most suitable method for our research topic.

3.4 Economic Valuation Methods

3.4.1 Introduction

Environmental protection is always an urgent issue. Though some progress has been made, many issues related to environmental protection are still in question. How much of the environment should be protected? How much will environmental protection cost and what benefits can be gained from environmental protection? In many respects, answers to these above questions remain ambiguous. Some work has been done on evaluating the costs of various environmental protection policies and projects, and it has been concluded that environmental protection generally comes at great cost. However, much less is known about the benefits of environmental protection policies and projects. Therefore, it is extremely important to quantify such benefits in order to make them more transparent and understandable. Consequently, the benefits of environmental protection can be considered seriously in any decision-making process (Garrod and Willis 1999). A number of techniques have been developed for valuing the benefits of environmental protection (see Fig. 3.4).

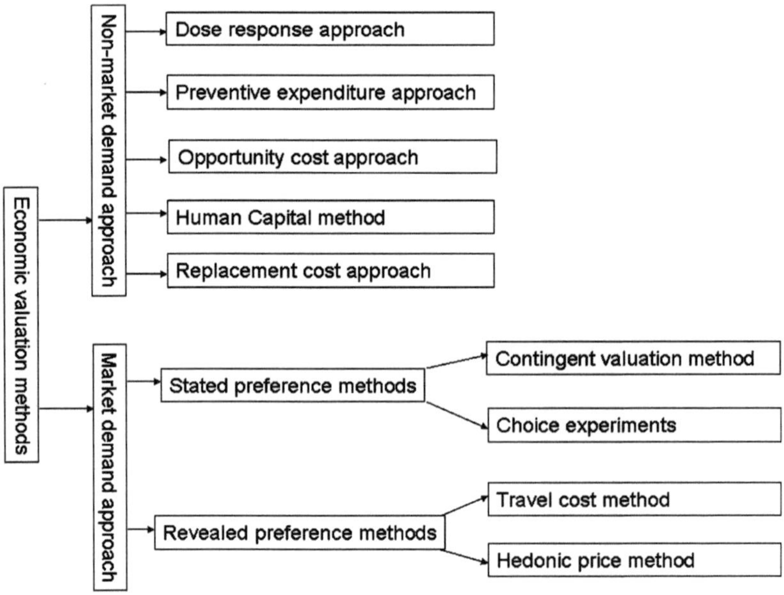

Fig. 3.4 Economic valuation methods (Source: OECD 1995)

According to Turner et al. (1994), valuation methods can be broadly divided into two categories: those that value a commodity via a demand curve, called market demand approaches; and those that do not value a commodity via a demand curve, called non-market demand approaches. Non-market demand approaches include the dose-response approach, preventive expenditure approach, opportunity cost approach, human capital method, and replacement cost approach (see Fig. 3.4). Market demand approaches are broadly classified into stated preference methods and revealed preference methods. The demand for environmental goods can be assessed by examining the purchases of related goods in the private market place. These may be complementary goods or substitute goods to the environmental goods. Alternatively, the demand for environmental goods can be measured by examining the individual's stated preference for these goods relative to their demand for other goods and services. Stated preference techniques ask individuals explicitly how much they value an environmental good (Garrod and Willis 1999). In the following, I explain the non-market demand approaches first.

3.4.2 *Non-market Demand Approaches*

Non-market demand approaches are designed to observe physical changes in the environment to estimate what differences they will make to goods and services, and then to estimate the market value of these changes. Non-market demand approaches include several methods as follows:

3.4.2.1 Dose-Response Approach

The dose-response approach seeks a relationship between environmental quality variables and the output level of a marketed commodity. For example, a dose-response approach can be used to estimate the cost of change in ground levels of ozone by seeking a relationship between ground levels of ozone and the output of crops. The cost of ozone change is equated with the loss of crop output, which can be valued at market prices (Garrod and Willis 1999).

3.4.2.2 Preventive Expenditure Approach

In the preventive expenditure approach, the value of the environment is inferred from what people are prepared to spend on preventing its degradation (Garrod and Willis 1999). For example, in order to reduce noise pollution, people are prepared to install double glazing.

3.4.2.3 Opportunity Cost Approach

Opportunity costs represent the value of the next best alternative forgone as the result of making a decision (McConnell and Brue 2005). The opportunity cost approach values the benefits of environmental protection in terms of what is being relinquished to achieve it (Garrod and Willis 1999). For example, a farmer encouraged to farm in an environmentally friendly way has to surrender some gains. The benefits of environmental protection can be estimated by the loss of crop output.

3.4.2.4 Human Capital Method

The human capital method links environmental change with human health. For example, emissions of particulate matter can affect human health. People would like to pay to avoid an increase in the risk of illness or death or to reduce the risk of illness or death. Their willingness to pay reflects the cost of air pollution caused by the emissions of particulate matter (OECD 1995).

3.4.2.5 Replacement Cost

Replacement costs refer to the amount it would cost to replace an asset at the current prices. The replacement cost approach values an environmental good by the cost incurred in restoring the environment to its original state after it has been damaged (Garrod and Willis 1999). For instance, the ecosystem service of a river can be valued by estimating the cost of restoring the polluted river to its original state.

Non-market demand approaches have been widely used by government to assess the cost of environmental impact. They are very straightforward and understandable. They are mainly applied in the fields, including to measure the effect of soil erosion on crop yields, the impact of environmental pollution on human health, and the effects of acid rain on crops, trees, and equipment.

Though non-market demand approaches are extensively used, they are not suitable for all situations. They are only appropriate for circumstances when environmental changes have a direct and observable effect on some goods or services that have market prices. The protection of YFPs impacts ecological services but ecological services do not have market prices. That is, non-market demand approaches are not suitable for evaluating porpoise protection. Therefore, the valuation of YFP can only be done by market demand approach.

3.4.3 Revealed Preference Methods

Revealed preference methods reveal the demand of environmental goods by studying the demand of complementary goods or substitute goods in the private market place. Revealed preference methods mainly include the travel cost method and the hedonic price method.

3.4.3.1 Travel Cost Method

The travel cost method (TCM) seeks to place a value on non-market environmental goods by using consumption behavior in related markets. Specifically, the costs of consuming the environmental goods or services are used as a proxy for price. These "consumption costs" for generating trips (e.g., to visit a national park) will include travel costs, entry fees, on-site expenditures, and expenditure on equipment necessary for consumption. Travel costs also include distance costs and time costs. The method assumes weak complementarities between environmental assets and consumption expenditures (Hanley and Spash 1993). That is, the consumption expenditures indicate the marginal utility of the environmental asset. The higher the consumption costs, the bigger the marginal utility of the environmental asset is. When no consumption takes place, the consumption expenditure is zero and, correspondingly, the marginal utility of the environmental asset is also zero.

The TCM can claim to be the oldest of the non-market valuation techniques, originating in a letter from the economist Harold Hotelling to the director of the US Park Service in 1947 (Hanley and Spash 1993). After that, Wood and Trice (1958) and Clawson and Knetsch (1966) gradually introduced the method into literature. TCM is predominantly used in outdoor recreation modeling, with fishing, hunting, boating, and forest visits among the most popular applications (Willis and Benson 1989).

In general, the TCM is suitable for accessible sites where people would like to expend significant time or other costs to travel. This method has been used to assess the use values of an urban museum (Martin 1994), of attending performances at the Royal Exchange Theatre in Manchester (Forrest et al. 2000), of visiting the historic city of St. Mary, Maryland (Poor and Smith 2004), and of four instances of cultural tourism in Spain (Bedate et al. 2004). Alberini and Longo (2006) combined the TCM with contingent behavior questions to estimate domestic visitors' use values for cultural heritage sites in Armenia.

Taking as an example, travel to Poyang Lake for the sightseeing of YFPs, the cost for generating a trip will include the travel costs such as for bus tickets, hotel fees, food costs, and opportunity costs caused by the time spent on the trip. The sum of these costs can be used as a proxy for the price of the YFP protection. The more tourists spend on the trip, the greater is the marginal utility of the YFP protection. Conversely, if people are not interested in such a trip and the trip does not happen, then the travel costs are zero, and the marginal utility of the YFP protection is consequently zero. Certainly, the utility of porpoise protection is not zero because of the option values and non-use values, even though nobody travels to the Poyang Lake for the sightseeing of YFPs. That is, the TCM can only estimate use values.

The TCM is applicable to the protection of YFPs in Poyang Lake. But porpoise-oriented tourism is so far underdeveloped. Matching equipment for watching YFPs has not been established in Poyang Lake. Few would like to travel to Poyang Lake for the sightseeing of YFPs at significant cost. Therefore, the TCM is not well-suited for the valuation of porpoise protection.

3.4.3.2 Hedonic Pricing Method

The hedonic pricing method (HPM) seeks to find a relationship between the levels of environmental services and the prices of the marketed goods on the proposition that an individual's utility for a good or service is based on the attributes it possesses. The hedonic pricing method identifies environmental service flows as elements of a vector of characteristics describing a marketed good, typically housing (Hanley and Spash 1993).

The first step in HP study is to decide which environmental quality variable is of interest, and then to ascertain whether sufficiently disaggregated, spatial data are available, along with housing price data. A hedonic price function can then be estimated, and then followed by an estimation of the demand curve (Freeman 1979).

For example, house prices (P_h) in the Poyang Lake region might depend on: site characteristics (S_i), such as the area of the house, traffic conditions, and whether a parking place is provided or not; neighborhood characteristics (N_j), such as crime rate and number of schools in the area; and environmental quality variables (Q_k),

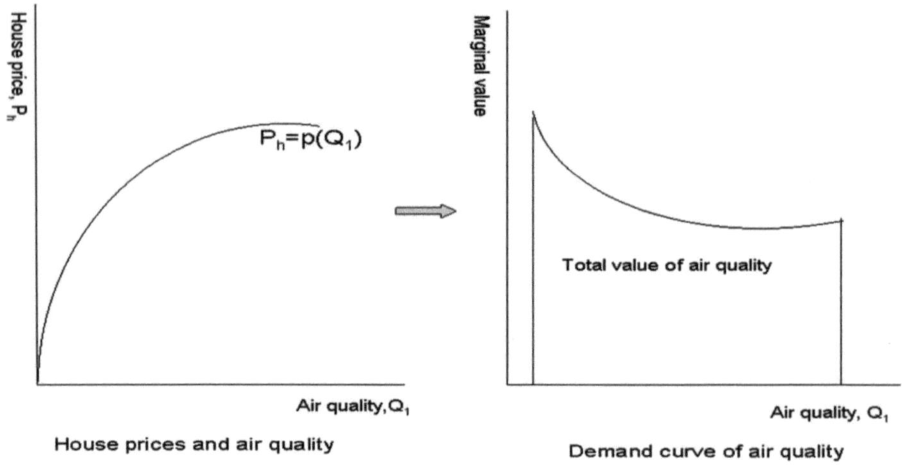

Fig. 3.5 Demand curve of air quality in housing prices (Source: Hanley and Splash 1993)

such as air quality, noise levels, surrounding landscape quality, and green coverage. Thus, the HP function of houses in Poyang Lake can be defined as:

$$P_h = P(S_i, N_j, Q_k) \bullet [i = 1 \ldots m, j = 1 \ldots n, k = 1 \ldots l]$$

Then the implicit price for air quality, Q_1, can be calculated as:

$$\delta P_h / \delta Q_1 P_h = P(S_i, N_j, Q_k).$$

The implicit price is a measure of valuing the marginal change of the air quality variable Q_1. On the basis of the equation calculating the implicit price, the demand curve for air quality can be estimated (see Fig. 3.5). With the demand curve of air quality, the total value of air quality can be aggregated.

The first studies on the HPM were published in the late 1960s and early 1970s (for example, Ridker and Henning 1967; Anderson and Crocker 1971). Since then, the application of the HPM to examine the impact of environmental factors on house prices has become widespread, such as environmental risk (Brookshire et al. 1985), landscape (Morales 1980; Anderson and Cordell 1985; Garrod and Willis 1992), water quality (Kirshner and Moore 1989), environmental protection (Frech and Lafferty 1984), urban amenity (Darling 1973; Wabe 1970; Correll et al. 1978), agricultural land values (Miranowski and Hammes 1984; Ervin and Mill 1985; Palmquist and Danielson 1989), pollution (Pearce and Markandya 1989; Harrison and Rubinfeld 1978), noise (Roskill Commission 1971; Nelson 1980, 1982), and social factors (Schafer 1979; Jud and Watts 1981; Forrest et al. 1996).

From the above studies, I can conclude that the HPM is appropriate when environmental quality is perceived as a factor that affects property values in an active and competitive property market, where market transactions are transparent. That is, environmental quality is closely linked with property values, such as for land and houses.

It happens that the protection of YFPs in Poyang Lake is not closely linked with any property values, such as for land or houses. Though the ecological quality of Poyang Lake can be improved at the same time that its porpoises are protected, housing prices around Poyang Lake are not affected by this because local residents do not consider porpoise protection as a factor impacting housing prices. Furthermore, the HPM can only estimate present environmental benefits provided to the local residents but cannot predict future benefits. Bequest values—the values for future generations—are an important part of the TEV of porpoise protection, and cannot be estimated by this method. The HPM cannot be used for valuing porpoise protection in Poyang Lake.

3.4.4 Stated Preference Methods

Stated preference methods value the environmental assets by asking people explicitly how much the environmental assets are worth in their opinion. There are two basic types of stated preference technique, the contingent valuation method and choice experiments.

3.4.4.1 Contingent Valuation Method

The Contingent Valuation Method (CVM) is a survey-based technique for the valuation of non-market resources. The method involves surveying to assess how much people would be willing to pay for specific environmental services (willingness to pay—WTP). In other cases, people are asked for the amount of compensation they would be willing to accept in order to give up specific environmental services (willingness to accept—WTA). It is called "contingent" valuation, because people are asked to state their willingness to pay, *contingent* on a particular hypothetical scenario and description of the environmental service (http://www.ecosystemvaluation.org/contingent_valuation.htm).

The CVM is an important valuation tool and has been widely applied by resources economists, following a great deal of empirical and theoretical refinements in the 1970s and 1980s. The contingent valuation survey was originally proposed by Ciriacy-Wantrup (1947) as a method of eliciting market values of non-market goods. It was first applied by Davis in 1963 to estimate the place of value hunters and tourists in a specific wilderness area (Davis 1963). Since then, the CVM has been widely applied in various fields, such as air quality (Wang and Mullahy 2006), biodiversity and recreation (Boxall et al. 1996; Ready and Navrud 2002; Mathieu et al. 2003; Jim and Chen 2006), wildlife (Loomis and Larson 1994; Langford et al. 2001; Jin et al. 2008; Loureiro and Ojea 2008), forest and wilderness areas (Adamowicz et al. 1998; Kniivilä 2006; Baral et al. 2008), ecosystem services (Loomis et al. 2000; Xu et al. 2003; Yang et al. 2008), sewerage and waste management (Jin et al. 2006). The protection of YFPs can be classified as wildlife conservation. Therefore, the CVM is also fit for the valuation of the protection of YFPs.

In the following, I briefly describe the procedure of CVM as well as the advantages and disadvantages of the method. The purpose of this section is to give an overview. Details will be provided in subsequent chapters.

1. Procedure of the contingent valuation study
   A CVM study can be split into six stages (Hanley and Spash 1993), which are here explained in detail:

   The first step is to set up a hypothetical market for the environmental service flow in question. For example, the Wuhan Baiji Conservation Foundation would like to contribute to the protection of YFPs in Poyang Lake by building a nature reserve there. The foundation itself has inadequate funding so wants to raise money from society by way of donations. Respondents may be told that the project is approved by the central government. They may also be told the location and area of the proposed nature reserve, and the expected effects of the project. Moreover, it should be made clear to respondents that the project cannot go ahead without adequate funds being raised. Besides information on all aspects of the hypothetical market, some basic information about the YFPs, such as their population, protection status, and an explanation of the need to protect them, will also be offered in a questionnaire. Together, the information constitutes the "framing" of the good.

   The second step is to obtain bids. Once the questionnaire is designed, the survey is administered. There are several survey methods available, such as face-to-face interview, web-based interview, telephone interview, mail interview, and group-based interview approaches (all the approaches will be explained in detail in Chap. 4). Individuals are asked to state their maximum WTP in order to gain an environmental improvement or to avoid an environmental deterioration. Or else, respondents are asked to indicate their minimum WTA for putting up with an environmental deterioration or forgoing an environmental improvement. There are quite a few elicitation formats for gaining WTP/WTA bids. Taking WTP as an example, WTP figure can be derived in several ways, including an open-ended question, a close-ended question, a dichotomous choice referendum, an iterative bidding game, and a payment card. These elicitation formats will also be explained in detail in the Chap. 4 regarding questionnaire development.

   The third step is to calculate the mean and median WTP/WTA. Once WTP/WTA bids have been gathered, mean and median bids are calculated. At this stage, it is very important to identify valid bids and to omit protest bids. Average bids are easily calculated when the payment-card, open-ended value, or bidding game approaches are used. However, if a dichotomous choice referendum approach is used, a Logit equation relating the probability of "yes" answers to each suggested amount must be estimated. The area under this curve gives a mean for WTP (Hanley and Spash 1993). Alternatively, the Cameron and James approach may be used to estimate a mean for WTP directly from a Logit or Probit equation (Cameron and James 1987; Cameron 1988).

   The fourth step is to estimate bid curves. With a series of WTP/WTA bid amounts, a bid curve can be estimated relative to a range of independent variables, such as income, age, education, and so on. For example, assuming that a series of

WTP bids gained on the protection project of YFPs, WTP bids might be regressed against income (Y), education (E), age (A), household members (M), gender (G), environmental attitude (J). The function can be defined as:

$$WTP_{\mathrm{i}} = \mathrm{f}\,(Y_i, E_i, A_i, M_i, G_i, J_i),$$

where i indexes respondents.

Estimation of bid curves against the socioeconomic variables is useful in aggregating results and for assessing the validity of the CV study.

The fifth step is to aggregate data. Aggregation refers to the process whereby the mean bid or bids are converted to a population total value figure. The figure should contain all the components of the value, such as use values, option values, and non-use values. The aggregation process involves three issues, including the choice of the relevant population, converting the sample mean value to a mean value of the total population, and the choice of an adequate time period. The choice of relevant population should be decided when constructing the sampling frame. When I convert the sample mean to the mean for the total population, I need to make sure that the sample mean value is representative of the total population, which is determined by the representativeness of the sample population that needs attaining in the sampling process. Then I can multiply the sample mean by the total population directly. The choice of time period depends on the setting within which the contingent valuation study is performed.

The final and sixth step is to evaluate the contingent valuation study. For example, does the survey result in a high proportion of protest bids? How well did the hypothetical market capture all aspects of the environmental good? Details on these design issues will be provided in the following chapters.

2. Advantages and disadvantages of CVM

The CVM is a technique with great potential utility and it is applicable to problems and circumstances that fall outside the scope of other methods. Such popularity attributes to its following advantages (see Ecosystem Valuation):

CVM is extremely flexible in that it can be used to estimate the economic value of almost anything. However, it is most capable at estimating values for the environmental goods and services that are easily identified and understood by respondents. For example, the protection of YFPs is very specific, and relates to the protection of other rare species, such as the giant panda. In the CV survey, an information folder will be offered to respondents, containing basic information about porpoise protection, such as population, distribution, protection status, why to protect them, and how to protect them. In addition, when respondents have any questions, they can ask surveyors. This is how porpoise protection can be understood by individuals.

CVM is the most widely accepted method for estimating TEV, including all types of use values and "passive use" values. The CVM can estimate use values, as well as existence values, option values, and bequest values. The "passive" use values account for most of the TEV of the protection of the YFPs. Therefore, the CVM is a good choice for evaluating the TEV of porpoise protection.

The results of CV studies are easily understandable. SPSS software can be applied to help analyze the survey results. The sample mean or median WTP or WTP values will ultimately be offered in the form of a monetary value. Furthermore, an aggregate value for the gross relevant population can be given as the TEV of a valued environmental good or service. Jin et al. (2008) estimated the household's WTP for black-faced spoonbill conservation in Macau, which ranges from 2.25 U.S. dollar to 38.56 U.S. dollar per household per month within 5 years. Yang et al. (2008) estimated the TEV of a constructed wetland program as 800,000 RMB. These results are easily grasped by people.

A great deal of research has been conducted on improving the methodology. For instance, how can the survey results be made more valid and reliable? How can some limitations of the method be eliminated or controlled? Mitchell and Carson (1989) suggested that the validity of the CV estimates should be judged, such as content validity, criterion validity, and construct validity, which will be detailed in a subsequent chapter. Additionally, some scientists, like Mazzanti (2001), Loomis (1994), Hagemann (1985) have applied the CVM to value the TEV of whales conservation, which offers a good model for our own research on the protection of YFPs, because YFPs and whales are both rare mammals of the Cetacean order.

On account of these advantages, the CVM is becoming more and more popular. However, the CVM is not perfect and it still has some disadvantages.

It should be noted that the method is very data-intensive, and the proper conduct of surveys is costly and time-consuming because of the extensive pretesting and survey work required. In addition, the design of surveys and interpretation of their responses require specialized skills.

Its greatest weakness is that it relies on people's views, rather than evidence of their market behavior. Many individuals are not familiar with the monetary values of environmental assets. More or less, it is challenging for respondents to give completely true statements. In consequence, some biases, such as the warm glow effect, embedding effect, ordering problem, strategic bias, and so forth, may rise in responses, although some of these can be controlled—if not eliminated—by appropriate survey design. These biases will be discussed and analyzed in the subsequent chapters. The CVM greatly relies on respondents' understanding of the environmental issue. This assumes a certain level of education and environmental awareness. These conditions are more likely to be satisfied for urgent local concerns than for national or global issues.

3.4.4.2 Choice Experiments

While the CVM concentrates on a particular scenario and provides information on preferences for the whole scenario, choice experiments can be used to examine the response of the individual to changes in the attributes of the scenario, as well as to the overall scenario. The choice experiment approach allows researchers to break down the relevant attributes of a situation and determine preferences over attributes, which increases the flexibility of the analysis (Garrod and Willis 1999). In a choice experiment, respondents are presented with a series of alternatives, differing in terms of attributes and levels, and asked to choose their most preferred (Hanley et al. 2001).

Empirically, choice experiments have been used in the market research and transport literatures (Green and Srinivasan 1990; Hensher 1994). Choice experiments have only recently been used in environmental valuation studies, such as recreation (Adamowicz et al. 1994, 1997; Boxall et al. 1996), forest and wilderness area (Adamowicz et al. 1998; Boxall and Macnab 2000; Bullock et al. 1998; Hanley et al. 1998; Blamey et al. 2000), wetland (Morrison et al. 1999), and water quality (Garrod and Willis 1999). In environmental valuation studies, the profiles used in choice experiments define the key attributes of an environmental good or service in question, and different profiles provide differential levels of these attributes and thus present different levels of utility to individuals (Garrod and Willis 1999). Respondents are asked to give the best choice among the profiles.

1. *Advantages*

   First, the method of choice experiment is better suited than the CVM in measuring the marginal value of changes in various characteristics of environmental programs. The method of choice experiment is particularly suited to deal with situations with multidimensional changes because the method is able to separately identify the value of individual attributes of a good or a program. In principle, the CVM can also be applied to estimate the value of the attributes of a program, for example, by including a series of CV scenarios in a questionnaire or by conducting a series of CV studies. However, it is a more costly and cumbersome alternative (Hanley et al. 2001).

   Secondly, choice experiments are more informative as respondents are given multiple opportunities to express their preferences for a valued good over a range of payment amounts. For example, if respondents are given eight choice pairs and a "do-nothing" option, they may respond to as many as 17 bid prices, including zero (Hanley et al. 2001).

   Thirdly, Choice experiment minimize some of the response difficulties found in the CVM, such as protest bids, strategic bias, and yea-saying, by providing a series of alternative choices from which WTP can be indirectly inferred (Hanley et al. 2001).

   While there are some advantages with the choice experiment, the application of choice experiment is still fairly limited in environmental studies because of some disadvantages.

2. *Disadvantages*

   The main disadvantage of the choice experiment lies with the cognitive difficulties associated with multiple complex choices between bundles with many attributes and levels. Experimental economists and psychologists have found ample evidence that there is a limit to how much information respondents can meaningfully absorb in making a decision (Hanley et al. 2001). For example, Mazzotta and Opaluch (1995) found that increased complexity leads to an increase in random errors.

In addition, since respondents are typically presented with a large number of choice sets, both learning and fatigue effects can occur, which may lead to apparently "irrational" choices (Tversky and Shafir 1992). Some respondents give repeated answers, which pose statistical problems in considering and modeling the correlation between responses (Adamowicz et al. 1998).

Moreover, the application of the choice experiment in estimating the total value of an environmental good or service is based on the assumption that the value of the whole is equal to the sum of the parts (Hanley et al. 1998). Some objections have been raised about this assumption. Research for London Underground and London buses, among others has shown clear evidence that values of whole bundles of improvements are valued less than the sum of the component values, all measured by using choice experiment (SDG 1999, 2000). Furthermore, Foster and Mourato (1999) found that the estimates from a choice experiment of the total value of the most inclusive public good were significantly larger than results obtained from a parallel CVM survey. However, the estimates of the total value of the less inclusive definition of the public good were significantly lower than the results from a CV survey.

Furthermore, it is more difficult for choice experiments to derive a value for a sequence of elements implemented by policies or projects, when compared to a CVM alternative. Hence, valuing the sequential provision of goods in multiattribute programs is probably better undertaken by the CVM (EFTEC 2001). The protection of YFPs is multi-attribution-able, such as water pollution control, noise pollution control, reduction of accidents, and so on. But porpoise population is increasing gradually. The effectiveness of the porpoise protection is sequential, greater and greater year by year. Therefore, it is better to apply the CVM to the porpoise protection projects than to apply the choice experiment.

Finally, welfare estimates obtained using choice experiment are sensitive to study design, which is common for all stated preference valuation methods. For example, the choice of attributes, the levels chosen to represent them, the way in which choices are relayed to respondents, are not neutral and may impact the values of estimates of consumers' surplus and marginal utilities. Hanley et al. (2000) found that changing the number of choice tasks respondents performed produced significant impacts on the model of preferences derived from their responses.

So far, I have explained all the popular valuation methods, analyzed the pros and cons of them, and summarized their application fields. Based on these characteristics, we can make our best choice for the porpoise protection project. The final choice is the CVM. Our arguments for this choice will be outlined in the following section.

3.4.5 Conclusion: Choice of Contingent Valuation Method as the Adequate Method for the Evaluation of the YFPs

The goal of the economic valuation is the protection of YFPs in Poyang Lake. The purpose is to gain the TEV of the porpoise protection project. According to the analysis of the TEV of porpoise protection, I can judge that the option value and the existence value of the porpoise protection are probably much more important than the use values. Besides, there are only very few YFPs living in the world and they are facing the threat of extinction. Some direct use values, such as edible meat, blubber, oil, and skin, cannot currently be developed. Therefore, I should pay special attention to the indirect use values, option values, and non-use values including the bequest values and existence values.

Some of the arguments about the pros and cons of the above-mentioned valuation techniques and their suitability to evaluate the YFPs have been explained before. In this section, I sum up the major arguments. Knowing the strengths and weakness of valuation methods, I can exclude some valuation techniques because of their unfeasibility.

1. The non-market demand approaches are designed to observe the effect of environmental change on some complementary and substitutive goods or services and then value the effects. Therefore, the affected goods or services must have market prices. The protection of YFPs produces effects on many aspects, such as water quality and the ecological services of Poyang Lake, none of which have market prices. Hence, non-market demand methods are not suitable for valuing porpoise protection project.
2. The TCM tries to place a value on non-market environmental goods by using consumption expenditure as a proxy for prices of environmental goods or services. As mentioned above, the TCM can only measure the use values but cannot measure the TEV of porpoise protection.
3. The HPM tries to find a relationship between environmental quality and some property values. The protection of YFPs in Poyang Lake does not affect any marketed goods directly, such as land and houses. That is, no relationship exists between porpoise conservation and some property values. Thus, the values of the porpoise protection cannot be valuated with HPM.
4. The above valuation approaches are abandoned because they share the problem of being unable to evaluate the TEV of porpoise protection. The choice experiments are a good method for measuring the TEV of environmental goods or services. However, the choice experiments are still not our best choice for two reasons: the protection of YFPs is a long-term process and the derived values are continuous and sequential, leading us to conclude choice experiments are not a good choice.
5. We now examine whether, and why, CVM is suitable for the valuation of TEV of the YFPs: the CVM is flexible enough that it can be applied to estimate the

economic value of anything, including the protection of YFPs. It is relatively easily understood by respondents, which is also an important criterion for choosing the method. This is particularly so given that the economic valuation of environmental assets is relatively new in China and residents are unfamiliar with the monetary value of environmental goods and services. The choice of the easier method will help to lower the complexity of the valuation study. Thus, CVM is the best choice for porpoise protection research.

The CVM, as one stated preference method, is based on welfare economics theory and the estimated results are consistent with welfare measures. WTP and WTA are instruments for realizing welfare measures. Therefore, I now turn to the theoretical foundation of the CVM.

3.5 Theoretical Basis of Contingent Valuation Method

The ultimate aim of a CVM study is to obtain an accurate estimate of the benefits or costs of a change in the provision of some public good, which can then be used in a benefit-cost analysis (Mitchell and Carson 1989). In order to achieve this, the survey must simultaneously meet the methodological imperatives of survey research and the requirements of economic theory. In order to meet the methodological imperatives, the basic scenario for the CVM is that it should be understandable and meaningful to the respondents and free of incentives, which could bias the results. In order to meet the requirements of economic theory, a survey must obtain the correct benefit measures for the good in the context of an appropriate hypothetical market setting. In this section, I review the theoretical underpinnings of the CVM, including welfare measures, willingness to pay, and willingness to accept. The methodological aspect of designing the questionnaire will be addressed in Chap. 4.

3.5.1 Welfare Measures

Welfare refers to the well-being of an individual, group, or economy (Stiegler 1976). In the context of the YFPs, welfare measures focus on the behavioral characteristics of the individual, which is the consumer. That is, the consumption characteristics are the research objects here.

The goal of consumption is assumed to be utility maximization, which is restrained by certain constraints, principally income and prices. A rational consumer will always choose the most preferred bundle of goods from a set of feasible alternatives (Hanley and Spash 1993). In Fig. 3.6, a consumer's best choice is given where an indifference curve is tangent to the budget constraint. In microeconomics, the indifference curve consists of consumption bundles, among which the consumer is indifferent. That is, all the consumption bundles on the indifference curve are equally

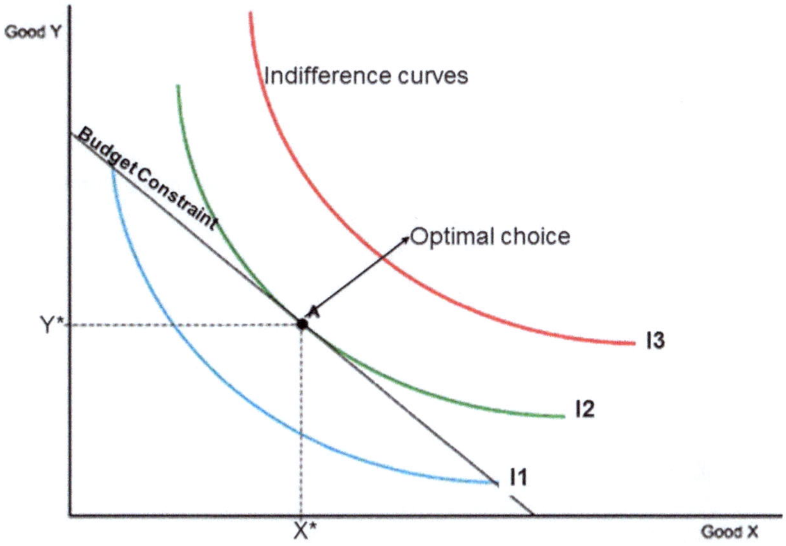

Fig. 3.6 Optimal choice, depending on indifference curve and budget constraint (Source: Varian 1987)

desirable. A budget constraint represents the combination of goods and services that a consumer can purchase given current prices and his income (Varian 1987).

3.5.1.1 Consumer's Surplus

Dupuit (1844) described a consumer's surplus as being the difference between the price that is actually paid when purchasing a commodity and the price the consumer would have been willing to pay. The WTP diminishes as more units of the commodity are consumed because the extra satisfaction derived from any good declines as more units of the commodity are consumed (Hanley and Spash 1993).

Marshall was concerned with finding the conditions under which a money measure of consumer welfare would equal the true utility surplus. He developed the association of consumer's surplus with the curvilinear triangle under the ordinary demand curve (Marshall 1890; Hanley and Spash 1993). Given a market-demand curve, as shown in Fig. 3.7, I can analyze a consumer's surplus. Demand curve DD' slopes down to the right because of the diminishing marginal utility of the good. When the market price is OC, the quantity demanded will be OH. There are some buyers who would have been willing to pay a higher price than OC, such as OC'. PR stands for the consumer surplus. The total consumer surplus is equal to CDA, which is the difference between what a consumer spends to buy the commodity (the area in Fig. 3.7: OCAH) and what he would have been willing to spend (the area in Fig. 3.7: ODAH). The Marshallian consumer surplus represents the gain in utility from consuming the good when the marginal utility of money remains constant.

Fig. 3.7 Marshallian consumer's surplus (Source: Hanley and Spash 1993)

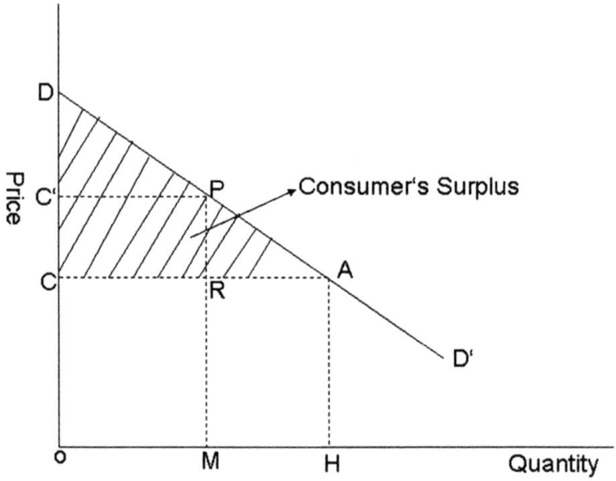

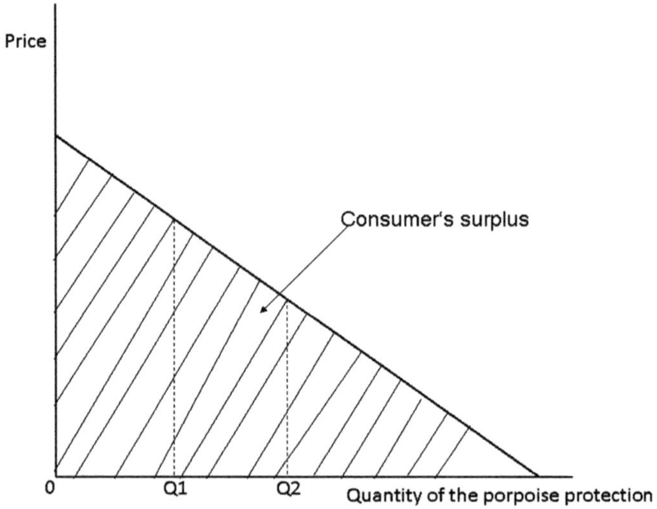

Fig. 3.8 Marshallian consumer's surplus for the public good (Source: Hanley and Spash 1993)

When the Marshallian consumer surplus is applied to the public good, such as the protection of YFPs, the figure will change into Fig. 3.8 because the market price of the public good is zero.

Unfortunately, the concept of the Marshallian consumer surplus has been shown to be problematic as a measure of the benefits resulting from price or quantity changes (Samuelson 1947; Silberberg and Suen 1978). These problems result from the fact that the ordinary demand curve does not hold the level of utility or satisfaction constant, but rather holds income constant (Mitchell and Carson 1989). In the

Marshallian consumer surplus, the marginal utility of the commodity is diminishing as the consumed quantity is increasing. There is a budget line, which indicates that the income does not change.

3.5.1.2 Hicksian Welfare Measures

In order to overcome the drawbacks of Marshallian consumer surplus, Hicks refined four measures of changes in consumer welfare resulting from a price change. These are compensating variation, equivalent variation, compensating surplus, and equivalent surplus.

Compensating variation is defined as the amount of income that must be taken away from a consumer (positive or negative) after an economic change to restore him to the original welfare level. In other words, the compensating variation is the income adjustment required to maintain the consumer at the utility level that occurred before price and income changes. Since the income change will compensate the consumer for the price change, it is called compensating variation in income (Varian 1987).

Porpoise protection is taken as an example here. Because porpoise protection is a public good, there is no market price for this. Therefore, a value is applied instead of a price, which supposes we are interested in the effect of an increase in the value of porpoise protection, named good X, on a particular consumer. The value rises from P_{X1} to P_{X2}. The individual has M Euro to allocate between X and Y, where Y represents all other goods, such as sand, and have price P_Y. P_Y and M do not have a subscripts because they will remain constant throughout the analysis.

The measure of consumer welfare changes resulting when a price changes—compensating variation, equivalent variation, compensating surplus, and equivalent surplus—can be derived in six steps:

Step 1: Find the Initial Equilibrium
The budget constraint before the value increases is the following:

$$BC_1 : M = P_{X1} * X_1 + P_Y * Y_1$$

Given the budget constraint, he will choose bundle 1 in Fig. 3.9, where the budget constraint is tangential to the indifference curve IC_1.

Step 2: Find the New Equilibrium Following Price Increase
When the value of porpoise protection X ascends to P_{X2}, the maximum consumption of the porpoise protection will reduce. The indifference demand curve will shift to the left and becomes IC_2. The budget constraint changes to the following:

$$BC_2 : M = P_{X2} * X_2 + P_Y * Y_2$$

The new budget constraint is labeled BC_2 in Fig. 3.10. The new optimal choice is bundle 2, where the new budget line BC_2 is tangential to the indifference demand curve IC_2. Bundle 2 is clearly worse off than bundle 1 because IC_2 is lower than IC_1.

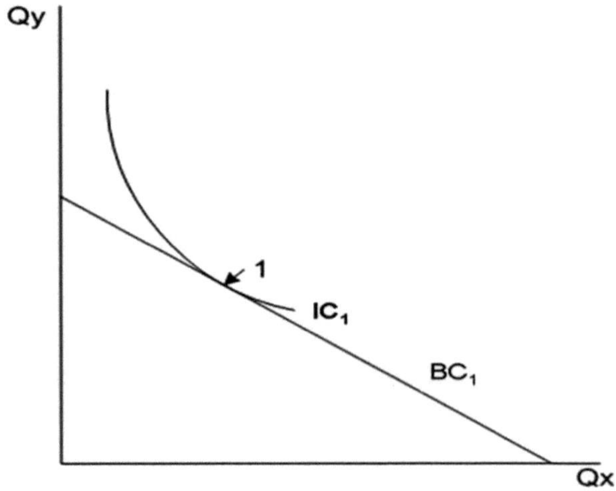

Fig. 3.9 Initial optimal choice before the value change (Source: Varian 1987)

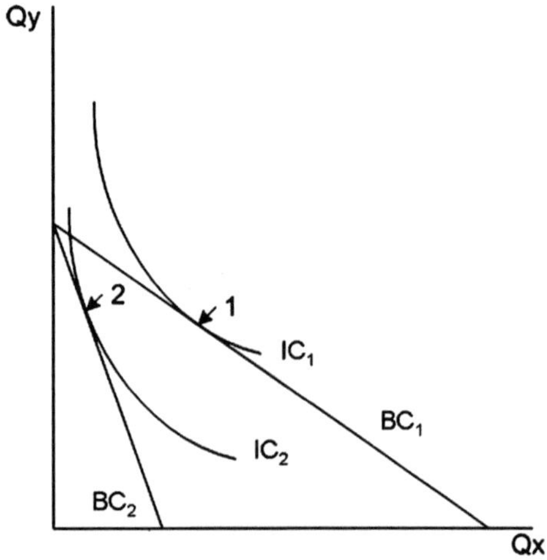

Fig. 3.10 New optimal choice after the value change (Source: Varian 1987)

Step 3: Finding the Compensating Variation

Now imagine that the government would like to increase the consumer's income after the price change. In principle, the consumer will have enough extra income that he can be held at his initial level of utility That is, he can reach his original indifference curve, IC_1, at bundle 3 where the budget constraint line BC_3 is tangential to the original indifference curve IC_1 (see Fig. 3.11).

With the increase in income, the consumer can be just as well-off after the value change as he was at the beginning. The vertical distance between the budget constraint line BC_2 and the parallel budget constraint line BC_3 represents the compensating variation. The budget constraint becomes:

$$BC_3 : M + CV = P_{X2} * X_3 + P_Y * Y_3$$

Fig. 3.11 Compensating variation for the value increase in the porpoise protection (Source: Varian 1987)

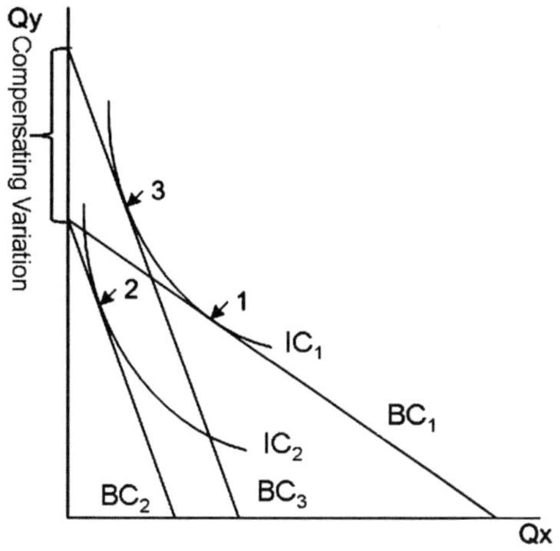

Step 4: Finding the Compensating Surplus
Compensating variation is used where the consumer is free to determine the consumed amount of the good. However, compensating surplus is applied where the consumer is required to consume certain amount of the commodity. *Compensating surplus* is defined as the compensating payment (For a case of increasing the porpoise population, it is consumer's maximum WTP; for a case of reducing the porpoise population, it is consumer's minimum WTA) that leaves an individual as well-off as before a certain environmental change, if he is forced to buy a given amount of the good (Mitchell and Carson 1989).

The situation is similar with that of compensation variation. The consumer can get some extra income from the government for the value change. But he is required to consume the amount X_2 of the porpoise protection, or good X. In principle, enough extra money can be offered to the consumer so that he can be as well-off after the value increase as he was before the value change. That is, he can reach his original indifference curve IC_1, at bundle 4 where the consumed amount of good X is still X_2 (see Fig. 3.12).

With the increased income, the consumer can be as well-off after the value increase as he was before the price change when he is required to consume the amount X_2 of the porpoise protection. The vertical distance between the budget constraint line BC_2 and the parallel budget constraint line BC_4 represents the compensating surplus for the value increase. Then, the budget constraint becomes:

$$BC_4 : M + CS = P_{X2} * X_2 + P_Y * Y_4$$

Step 5: Finding the Equivalent Variation
In co ntrast to the compensating variation, the *equivalent variation* is defined as the amount of income that must be given to a consumer (positive or negative) before an economic change to leave him as well-off as he would be after the price change. This is called the equivalent variation in income, since it is the income change that is the equivalent change in terms of price change in utility (Varian 1987).

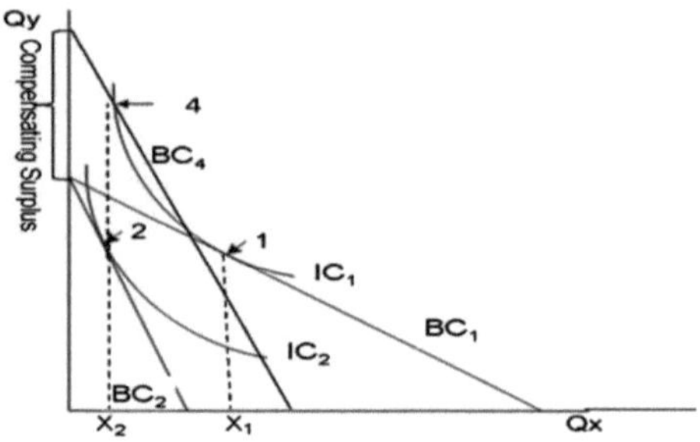

Fig. 3.12 Compensating surplus for the value increase in the porpoise protection (Source: Hanley and Spash 1993)

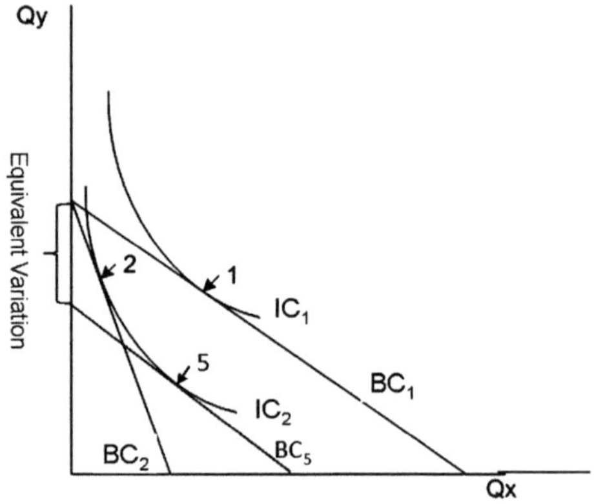

Fig. 3.13 Equivalent variation for the value increase in the porpoise protection (Source: Varian 1987)

Now imagine the government wants to decrease the consumer's income by way of tax before the value increase. In principle, the consumer's well-being is as good with the income reduction as it would be after the value increase. That is, the consumer will reach the new indifference curve IC_2, at bundle 5 where the budget constraint line BC_5 is tangential to the new indifference demand curve IC_2 (see Fig. 3.13). The budget constraint line BC_5 is parallel with the original budget constraint line BC_1.

With the reduction in income the consumer is as well-off without the value change as before it had occurred. The vertical distance between the original budget line BC_1 and the parallel budget line BC_5 stands for the equivalent variation in income. Then the budget constraint becomes:

$$BC_5 : M - EV = P_{X2} * X_5 + P_Y * Y_5$$

Step 6: Finding the Equivalent Surplus

Equivalent Surplus is defined as the amount of income that must be given to a consumer (positive or negative) before an economic change to leave him as well-off as he would be after the price change. The consumer is forced to consume a given amount of the good (Mitchell and Carson 1989).

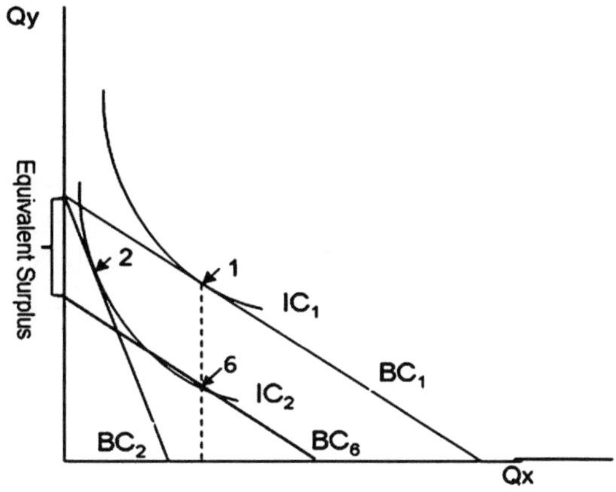

Fig. 3.14 Equivalent surplus for the value increase in the porpoise protection (Source: Hanley and Splash 1933)

Suppose that the government would like to reduce the consumer's income by way of tax before the value increase. The consumer is still required to buy the given amount X_1 of the porpoise protection, good X. In this situation, the consumer with the decreased income can be as well-off without the value change as it had occurred. That is, the consumer will reach the indifference curve IC_2, at bundle 6 where the consumed amount of the porpoise protection is still X_1 (see Fig. 3.14).

With the reduction in income, the consumer can be as well-off before the value change as it would be after the value increase when he still consumes the required amount X_1 of the porpoise protection. The vertical distance between the original budget line BC_1 and the parallel budget line BC_6 stands for the equivalent surplus in income. Then the budget constraint becomes:

$$BC_6 : M - ES = P_{X2} * X_1 + P_Y * Y_6$$

In summary, the compensating variation and compensating surplus mean the income adjustment *after* the price or value change while the equivalent variation and equivalent surplus refer to the income change *before* the price or value change. In the situations of compensating variation and equivalent variation, consumers are free to determine the consumption amount while consumers are forced to consume the given amount of certain commodity in the circumstances of compensating surplus and equivalent surplus. The four Hicksian welfare measures can measure the consumer's utility when his income and price change.

In the CVM study, income is an important factor that affects consumers' willingness to pay or willingness to accept. Therefore, Hicksian welfare measures are suitable for the CVM study. The choice of welfare measures depends on the study scenario. In the following, I will introduce two other important methods of measuring a person's utility, such as WTP and WTA.

3.5.2 Willingness to Pay

WTP is the amount of money an agent would be willing to give up to obtain a change and still be as well-off as with his previous entitlement (Mitchell and Carson 1989). For example, respondents would be willing to give up some money in order to obtain an increase in porpoise population or in order to avoid a decrease in porpoise population. The respondent's well-being is the same as that before giving up the money.

3.5.3 Willingness to Accept

WTA is the amount of money which would have to be given to an agent, with a specified entitlement, to forgo a change and still be as well off as if the change had occurred (Mitchell and Carson 1989). For example, people would be willing to accept some money as compensation for the loss of the porpoise population or for giving up an increase in the porpoise population. Then they are as well-off as without the porpoise population or with the present porpoise population.

The relationship among Hicksian welfare measures, WTP and WTA is summarized as: for a price increase, the equivalent variation and equivalent surplus measure the consumer's WTP to avoid the change while the compensation variation and compensation surplus measure the individual's WTA compensation to tolerate the change. Alternatively, for a price decrease, the equivalent variation and equivalent surplus measure the minimum amount the consumer is willing to accept to forgo the price change, while the compensation variation and compensation surplus measure the consumer's WTP to secure the change (Hanemann 1999). By understanding the relationship between these welfare measures, we can make correct choices according to the specific valuation scenario. The utility gained from the protection of porpoises can be measured and calculated with the welfare measure.

3.6 Challenges Faced by Contingent Valuation

By comparing various valuation methods derived from economic valuation literature, we conclude that the CVM is the best choice for our study. By analyzing the welfare measures that support the CVM study, we can decide the most appropriate welfare measure for valuing porpoise protection. That said, we still cannot ensure the valuation results will be most accurate because of some disadvantages of the CVM itself. Having described the theoretical basis of the CVM above (Sect. 3.5), we now turn to some challenges and disadvantages of this method and draw attention to open research questions that will be addressed in the subsequent chapters of this study.

3.6.1 Overview of Main Challenges

CVM has become one of the most widely used non-market valuation techniques of recent decades. The prominent advantage of the CV technique is in estimating the TEV, including non-use values. However, there is much controversy over whether this method adequately measures people's WTP for environmental assets.

First, the CVM assumes that respondents understand the good in question and will reveal their preferences in the CVM market just as they would in a real market. A frequent claim (e.g., Desvousges et al. 1993) is that familiarity with a good is a necessary prerequisite for providing "meaningful" responses to valuation questions. However, most people are unfamiliar with placing monetary values on environmental goods and services. Therefore, we have to make efforts to design the information folder and questionnaire in order to make them comprehensive, simple, and easily understood. In addition, we choose the group-based approach in order to ensure respondents are informed relatively quickly. This allows respondents to become familiar with the environmental asset, namely the protection of YFPs. Under this precondition, we hope that respondents will give true statements.

Secondly, respondents may make associations among environmental goods that the researcher had not intended. For example, when respondents are asked for their WTP for the protection of YFPs, they expressed more concern about water pollution controls in Poyang Lake and its relevance to fishery. Their willingness to pay is based on the increase in fish stocks fishery by lower levels of water pollution. Some respondents value what is not aimed at, and this is worth much attention.

Thirdly, the payment question can be phrased as "How much are you willing to pay at most in order to gain the environmental asset?" or "How little are you willing to accept as compensation for losing the environmental asset." In theory, the WTP and WTA results should be identical. In fact, it has been found that WTP and WTA estimates differed substantially in some CVM studies (Hanemann 1991; Coursey et al. 1987; Knetsch and Sinden 1984; Harless 1989). The key implication of this divergence for applied policy work is that property rights can have a substantial influence on the magnitude of the welfare measure, especially when considering a reduction in the environmental service (Knetsch 1990). It has been proven that the difference between WTP and WTA estimates exist and the difference cannot be eliminated. More research on it only proves repeatedly that WTP differs from WTA. Therefore, our own research does not focus on the point. I do not expect any progress from focusing on this point.

Fourthly, there are some effects on the estimation results, such as ordering effect, embedding effect, and warm glow effect. *Ordering effect* refers to the phenomenon that people's stated WTP for certain object depends on where it is placed on a list of the objects being valued (Randall et al. 1981). In our own research, the protection of YFPs is the only object to be valued. Hence, there is no ordering effect. *Embedding effect* denotes the phenomenon that the respondent's stated similar WTP amounts on the one part of an environmental asset and the whole asset (Hoehn and Randall 1989). It is estimated that people would state similar WTP amounts on the protection of YFPs in Poyang Lake and the whole Yangtze River system. However, our research only concentrates on porpoise protection in Poyang Lake so that the embedding effect is possibly identified in our research. *Warm glow*, termed by Becker (1974), recognizes that individuals also derive utility from the act of giving through the associated values of social approbation, prestige, or moral satisfaction. That is, respondents express their feelings about the scenario or the valuation study itself rather than stating a value for the environmental good. For instance, respondents may state a positive WTP for porpoise protection because they feel good about the act of giving for a social good, although they believe that the good itself unimportant. The warm glow effect can only be identified but cannot be reduced or prevented. Therefore, I do not treat their aspect in more detail.

Fifthly, a *strategic bias* arises when the respondent provides a biased answer in order to influence a particular outcome, which has long been a concern of economists dealing with public good issues (Samuelson 1954). For example, some respondents who support porpoise protection may attempt to provide an answer of higher value than their true statement in order to ensure the high value of porpoise protection and strengthen the possibility that more porpoise protection projects are carried out. In this case, it is recommended that the survey scenario should be made as hypothetical as possible in order to avoid strategic behavior.

Sixthly, *budget constraint* is an important factor for affecting respondents' WTP. Some research results have indicated that WTP is positively correlated with income (Mataria et al. 2006; Keyserlingk and Rhodes 2007; Pavlova et al. 2004). Income differences exist among respondents and are unavoidable. Hence, the income effect cannot be eliminated. It is possible that WTP is distorted by budget constraints or ATP. Nevertheless, our research confirms the significant role played by the income effect (see Chap. 5).

Finally, the *elicitation format* and *payment vehicle* are also related to the true preference statement, which will be explained in the chapter on questionnaire development (see Chap. 4). In order to check these possible error sources, we need to test the validity and reliability of the CVM study (see Chap. 5). Among the error sources, *distance decay* will be paid special attention in our research because the effect is of great importance in aggregating the TEV of the porpoise protection project in Poyang Lake.

In the following, we take the last aspect—the distance effect—and address it in more detail. This is because there are some gaps in the literature. In addition, we expect that distance decay to play a major role in valuing the protection of YFPs.

3.6.2 Distance Decay as a Specific Challenge

3.6.2.1 Concept

People tend to discount their fear of (desire for) an object the further they are away from it—negative (positive) geographic discounting (Hannon 1994). As the distance from the proposed object of concern increases, the level of concern on it declines.

Fig. 3.15 Study sites in China (Source: http://www.world-atlas.biz/china.htm, plus own presentation)

In CV studies, a large problem is that CV judgments are sometimes insensitive to the scope of the good provided (Baron and Greene 1996). A predetermined geographic boundary for the sampling frame is utilized in most existing CV studies (Pate and Loomis 1997). When distance is not explicitly incorporated into the CV results analyses, it is implicitly assumed to have a constant effect across the study region. As a result, values are underestimated when too small a boundary is placed on the geographic extent of influence. On the contrary, values are overstated when the sample estimate is extrapolated over too large a region (Sutherland and Walsh 1985).

In CV studies, "distance-decay" is used to refer to the phenomenon whereby the mean value placed on a given environmental improvement falls, the further away an individual lives from the improvement (Hanley et al. 2003). For example, in order to check the distance effect on the preservation values of the YFPs in Poyang Lake, survey is supposed to be made in three cities, including Nanchang, Guangzhou, and Beijing (see Fig. 3.15). Nanchang is located in the Poyang Lake region. Guangzhou is closer to Nanchang than Beijing. According to the theoretical expectation, the mean WTP in Nanchang should be highest and the mean WTP in Beijing should be smallest. Mean WTP in Guangzhou should be bigger than that in Beijing and smaller than that in Nanchang.

In the following section, I will give an overview of the literature on distance decay. The goal is (1) to deepen the understanding about the knowledge on distance decay, and (2) to develop my own hypothesis about distance decay in the case of the YFPs.

3.6.2.2 Literature Review

All the relevant literature and corresponding results are listed in Table 3.4. In summary, most existing studies focus on distance effect on *use values*. Distance effects play a *significant and negative* role on use values of some environmental goods, like water quality improvement (Georgiou et al. 2000), preservation of peat meadow land (Brouwer and Slangen 1998), recreational hunting use (Boxall et al. 1996), preservation of heather moorland and rough grassland (Hanley et al. 2001), and land conservation easements (Cho et al. 2008). Bateman et al. (2006) stated that there was a *negative but insignificant* relation between distance and WTP for improving water quality. However, Pate and Loomis (1997) concluded that distance *did not play a role* in the mainly use-values driven Salmon program. In addition, Racevskis (2005) stated that WTP for forest easement programs *increase* as the distance from the forest *increases* because individuals with biocentric attitudes, who participate in a combination of hunting, snowmobiling, and other recreational activities and live farther away from the forest, have higher WTP.

With respect to *non-use values*, Sutherland and Walsh (1985) concluded that distance effects played a *negative* but *insignificant* role on WTP for option, existence, and bequest values of water quality. Bateman et al. (2002b) found a *national distance decay effect* on the non-use values for reductions in acidity in Scottish mountain. Concu (2005) found that the non-use values of bushland in Kings Park were *distance-independent*.

There are some studies on distance effects for TEV of some environmental goods, which do not distinguish use values from non-use values. For example, Pate and Loomis (1997) concluded that distance played a *significant and negative* role for wetlands improvement and contamination control. At the same time, Washington was willing to pay *more* than Oregon for a wetlands improvement program and contamination control, which *countered to* the distance concept because of substitution effects caused by wetlands in Oregon. Bateman and Langford (1997) stated that per annum and lump sum WTP for a national park *decreased markedly* as distance *increased*. Moran (1999) estimated that WTP for a national park *went to zero* at a distance of 214 km. Hanley et al. (2003) concluded that distance-decay effects exist on WTP *for the River Mimram* improvements, but *not* on WTP for *all Thames region rivers*. In addition, stronger distance decay existed for *use values* than for *non-use values* for the River Mimram.

Table 3.4 Summary of studies on distance decay

Author	Year	Research object	Survey range	Types of use values	Results
Sutherland and Walsh	1985	Water quality in the Flathead river drainage system, Montana	640 miles	Option existence and bequest demands	The distance coefficients for the option, existence, and bequest equations respectively are of the proper sign but none is significant at 0.10 level; The distance coefficients are negative for each of preservation values, indicating that WTP for option value decreases at a rate of 1.5 ¢ per mile compared to decreases at a rate of 5.4¢ and 6.3 per mile for existence and bequest value.
Boxall et al.	1996	The effects of forest management practices on recreational moose hunting values	0–350 km	Use value (hunting)	Distance played a negative and significant role in determining WTP.
Loomis	1996	Removing two dams and restoring salmon populations in Elwha river	Nationwide in the USA	Mainly use values (recreational opportunities, hydroelectricity, habitat for fish, water supply and a source of jobs)	WTP dropped from about $ 78 per household near Seattle in Washington State to around $ 60 per household on the east coast of the US, nearly 3,000 miles away.

(continued)

Table 3.4 (continued)

Author	Year	Research object	Survey range	Types of use values	Results
Pate and Loomis	1997	Wetlands and contamination control program in the San Joaquin Valley, California; Salmon improvement program	0–1, 134 miles	Total values; Mainly use values for the salmon program	The coefficient on distance was significantly negative for the wetlands improvement model and the contamination control model; Distance did not play a role in the determination of an individual's WTP for the Salmon improvement program Washington is willing to pay more than Oregon for wetland improvement program and contamination control program, which seemed to counter to the distance concept. That is, WTP goes up with distance.
Brouwer and Slangen	1998	Preservation of peat meadow land	Alblasserwaard	Total values (agricultural use, habitat for meadow birds, rich vegetation)	Distance effect is negative for WTP for preserving peat meadow land. South Holland respondents appear to be WTP 18 cents per year less for each kilometer they live further away from the Alblasserwaard.
Bateman and Langford	1997	Protecting a national park named Norfolk Broads from the saline flooding	0 km to England, all of Wales and Scotland up to the southern edge of the Highlands	Total values	Per annum and lump sum WTP bids decreased markedly as distance from the national park increased
Moran	1999	The same as above	The same as above	The same as above	Mean WTP is predicted to go to zero at a critical distance of 214 km.

Georiou et al.	2000	Water quality improvements for the River Tame in Birmingham	Birmingham area	Use values (paddling, fishery, swimming, canoeing)	WTP declined to zero at a distance of 16 miles (for a small improvement) and 36 miles (for a big improvement)
Bateman et al.	2002	Reductions in acidity in Scottish mountain lochs across the Scotland-England border	Scotland-England	Non-use values	A national distance effect for the non-use values for reductions in acidity in Scottish mountain.
Hanley et al.	2003	Improving low flow conditions in River Mimram and all rivers in Thames region.	0–150 km	Use and non-use values	Distance-decay effects exist for improvements to the Mimram for both users and non-users; No significant distance effect exists for the improvements to all Thames regions rivers; A stronger distance-decay exists for use values than for non-use values for the Mimram.
Hanley et al.	2001	Preventing changes in heather moorland and rough grassland in Scotland	0–50 miles	Total values (landscape, habitat for some animals)	WTP for the heather moorland area is supposed to go to zero at a distance of 25 miles; WTP for the rough grassland is supposed to go to zero at a distance of 30 miles.

(continued)

Table 3.4 (continued)

Author	Year	Research object	Survey range	Types of use values	Results
Bateman et al.	2006	Improving water quality in the River Tame	Birmingham area	Mainly use values (fishing, plants and wildlife, boating and swimming)	There was a negative relation between increasing distance (from respondents' permanent address to the River tame) and WTP, but the coefficient is insignificant.
Racevskis	2005	Management of forest easement program in Michigan's Upper Peninsula	Michigan State 0–503 miles	Mainly use values (a source of job, recreation and tourism, habitat for migratory forest songbirds and deer)	Distance had a positive and significant effect on the probability of voting "yes" for the easement program; WTP for easement program increases as distance from the resource increases; Nonusers living farther from the resource have lower WTP than those living closer to the resource; Certain types of users, like hunting and snowmobiling, would like to pay more with increasing distance to the resource.
Concu	2006	Improving the bushland in Kings Park of Perth	0–>700 km	Total values	No distance effects were recorded for the WTP for eradicating weed; WTP for fire prevention decreases with distance and increases again; Reducing accessibility to Kings Park bushland concerns less residents living far away from the park. Distance played a negative role in WTP for supporting the preferred management strategy.

Concu	2005	Improving the bushland in Kings Park of Perth	0–>700 km	Non-use values	Non use values are found to be distance-independent.
Cho et al.	2008	Land conservation easements in Macon County, New York City	0–28.4 km	Use values	A resident furthest from the poorer quality stream (28.4 km) is 14.0% and 16.8% more likely to choose zero WTP and protest, respectively than a resident closest to the poorer quality stream (0.4 km); A resident closest to the poorer quality stream is 30.8% more likely to choose positive WTP than residents at the sample's upper limit of distance to poorer quality streams.

So far, it can be concluded that distance effects indeed exist for both use values and non-use values of some environmental goods. Also, the distance effects work differently due to the characteristics of environmental goods and the types of values. Distance-decay effects exist significantly for some values or some environmental goods but are insignificant for other values or environmental goods. At the same time, some environmental goods are immune to distance decay, as are some values of certain environmental goods. Now, I turn to the research on distance effects on porpoise protection in Poyang Lake region—China. First of all, I would like to explain why distance effects are researched in my study.

3.6.2.3 Importance of Distance Effects Research

There are several reasons to investigate distance effects on environmental preferences. First, distance affects the use of environmental goods, information, and substitution possibilities that, in turn, affect preferences (Sutherland and Walsh 1985). For example, individuals have different possibilities and then hold different preferences on the values of porpoise protection. It is easier for residents in the Poyang Lake region to go sightseeing than people outside the region. Therefore, local residents may go porpoise sightseeing more frequently and have greater preference for them than people living further away.

Second, aggregation of mean estimates is generally guided by political/administrative criteria. Benefits are assumed to differ from zero within given political boundaries and to be nil outside. Identifying distance effects would provide an empirical validation to this criterion (Loomis 1996). For instance, the role in maintaining biodiversity by the YFPs, as unique freshwater mammals, benefits not only local residents in the Poyang Lake region but also people around China, if not throughout the world. For this reason, aggregation of the mean WTP for the porpoises should not be confined to the administrative boundary of the Poyang Lake region. In terms of deciding whose WTP should be included in the aggregation, distance effects research can help determine the benefits gained by individuals.

Thirdly, since benefit transfer uses sample (or population) characteristics to transfer estimates from population A to population B, assessing the effect of distance would be helpful in benefit transfer applications (Bateman et al. 1999; Jiang et al. 2005). Assessing distance effects can help capture the spatial complexity of the factors determining the benefit transfer functions and validify benefit transfer studies. For example, Bateman et al. (1999) addressed the spatial complexity by capturing travel time variable, considering the availability and quality of roads and then allowing for the spatial distribution of potential visitors, which were all directly related with distance.

Fourthly, on a policy level, investigating distance effects can provide useful information regarding the appropriate form of funding for environmental projects (local, state, or federal) (Concu 2006). When the boundary—on which WTP goes to zero—is determined, it is also clear at which level the government should be responsible for funding porpoise protection. At present, the funding sources for porpoise protection come mainly from the state government. Detection of distance effects can tell whether it is reasonable that only the state government appropriate funds for porpoise protection.

In conclusion, as a rare species unique for living in the Yangtze River, it is very important to conceptualize who benefits from the YFP's conservation. Therefore, it is necessary to test the distance effects on the TEV of porpoise conservation.

3.7 Conclusion

Economic valuation is used to assess the economic value of the environment. Five advantages of monetary valuation, including comprehensibility, de-emotionalism, measurability, capability of internalization, and amendment of gross domestic product—GDP, determine the necessity of making a monetary valuation of environmental goods. It is very important to provide an economic valuation of porpoise protection because it is at risk of market failure. The protection of porpoises is a public good and its property rights are not clearly entitled. Porpoise protection produces some externalities that cannot be internalized into the cost-benefit analysis of the project.

Porpoise protection is essential due to the great TEV of the YFPs, including use values, such as sightseeing and maintaining biodiversity; option values; and non-use values. Non-use values include bequest value, altruistic value ,and existence value. CVM is selected for valuing the TEV of porpoise protection because it can estimate both use values and non-use values. Hicksian welfare measures, including compensating variation, compensating surplus, equivalent variation, and equivalent surplus, are the theoretical basis of CVM. WTP and WTA are applied in the CVM study to measure the consumers' utility. Though CVM is widely applied in various fields, the method is still presented with challenges, such as warm glow, strategic bias, embedding effect, and distance effect. I will focus on the distance effects on porpoise conservation because it is essential to conceptualize who benefits from porpoise conservation.

For CVM, it is crucial to establish a hypothetical market in order to realize the transaction. Because environmental goods are not always familiar with residents, it is essential that the environmental goods are introduced through the preparation and provision of an information folder. It is also important to pose some questions designed to gauge respondents' attitudes. This will account for a great part of the method in developing a complete questionnaire. An appropriate and reliable survey method should be chosen for finishing the questionnaire. In the subsequent chapters, I turn to questionnaire development and selection of survey method in detail.

References

Adamowicz, W., Louviere, J., & Williams, M. (1994). Combining revealed and stated preference methods for valuing environmental amenities. *Journal of Environmental Economics and Management, 26*, 271–292.

Adamowicz, W., Swait, J., Boxall, P., Louviere, J., & Williams, M. (1997). Perceptions versus objective measures of environmental quality in combined revealed and stated preference models of environmental valuation. *Journal of Environmental Economics and Management, 32*, 65–84.

Adamowicz, W., Boxall, P., Williams, M., & Louviere, J. (1998). Stated preference approaches for measuring passive use values: Choice experiments and contingent valuation. *American Journal of Agricultural Economics, 80*, 64–75.

Alberini, A., & Longo, A. (2006). Combining the travel cost and contingent behavior methods to value cultural heritage sites: Evidence from Armenia. *Journal of Cultural Economics, 30*, 287–304.
Alchian, A. (1965). Some economics of property rights. *Il Politico, 30*, 816–829.
Anderson, L. M., & Cordell, H. K. (1985). Residential property values improve by landscaping with trees. *Southern Journal of Applied Forestry, 9*, 162–166.
Anderson, R. J., & Crocker, T. D. (1971). Air pollution and residential property values. *Urban Studies, 8*, 171–180.
Baral, N., Stern, M. J., & Bhattarai, R. (2008). Contingent valuation of ecotourism in Annapurna conservation area, Nepal: Implications for sustainable park finance and local development. *Ecological Economics, 66*, 218–227.
Baron, J., & Greene, J. (1996). Determinants of insensitivity to quantity in valuation of public goods. *Journal of Experimental Psychology. Applied, 2*, 107–125.
Bateman, I. J., & Langford, I. H. (1997). Non-users' willingness to pay for a national park: An application and critique of the contingent valuation method. *Regional Studies, 31*, 571–582.
Bateman, I. J., Lovett, A. A., & Brainard, J. S. (1999). Developing a methodology for benefit transfers using geographical information systems: Modelling demand for woodland recreation. *Regional Studies, 33*, 191–205.
Bateman, I. J., Carson, R. T., Day, B., Hanemann, M., Hanley, N., Hett, T., Jones-Lee, M., Loomes, G., Mourato, S., Özdemiroglu, E., Pearce, D., Sugden, R., & Swanson, J. (2002a). *Economic valuation with stated preference techniques: A manual*. Cheltenham: Edward Elgar.
Bateman, I., Cooper, P., Poe, G., Navrud, S., Ready, R., & Vossler, C. (2002b). *Scope sensitivity in passive use values: Evidence from experimental and survey studies*. Paper presented to World Congress on Environmental and Resource Economics, Monterey, CA.
Bateman, I., Jones, A., Jude, S., & Day, B. (2006). *Reducing gains/loss asymmetry: A Virtual Reality Choice Experiment (VRCE) valuing land use change* (CSERGE Working Paper EDM 06-16). Retrieved from www.uea.ac.uk/env/cserge/pub/wp/edm/edm_2006_16.htm – 8 k.
Bator, F. M. (1958). The anatomy of market failure. *Quarterly Journal of Economics, 72*, 351–379.
Becker, G. (1974). A theory of social interactions. *Journal of Political Economy, 82*, 1063–1094.
Bedate, A., Herrero, L. C., & Sanz, J. A. (2004). Economic valuation of the cultural heritage: Application to four case studies in Spain. *Journal of Cultural Heritage, 5*, 101–111.
Berkes, F. (Ed.). (1989). *Common property resources: Ecology and community-based sustainable development*. London: Belhaven.
Blamey, R. K., Bennett, J. W., Louviere, J. J., Morrison, M. D., & Rolfe, J. (2000). A test of policy labels in environmental choice modeling studies. *Ecological Economics, 32*, 269–286.
Boxall, P. C., & Macnab, B. (2000). Exploring the preferences of wildlife recreationists for features of boreal forest management: A choice experiment approach. *Canadian Journal of Forestry Research, 30*, 1931–1941.
Boxall, P. C., Adamowicz, W., Swait, J., Williams, M., & Louviere, J. (1996). A comparison of stated preference methods for environmental valuation. *Ecological Economics, 18*, 243–253.
Bromley, D. W. (1991). The commons, common property and environmental policy. *Environmental and Resource Economics, 2*, 1–17.
Brookshire, D. S., Thayer, M. A., Tschirhart, J., & Schulze, W. D. (1985). A test of the expected utility model: Evidence from earthquake risks. *Journal of Political Economy, 93*, 369–389.
Brouwer, R., & Slangen, L. H. G. (1998). Contingent valuation of public benefits of agricultural wildlife management: The case of Dutch peat meadow land. *European Review of Agricultural Economics, 25*, 53–72.
Buchanan, J. M. (1965). An economic theory of clubs. *Economica, 32*, 1–14.
Buchanan, J. M., & Stubblebine, W. C. (1962). Externalities. *Economica, 29*, 371–384.
Bullock, C., Elston, D. A., & Chalmers, N. A. (1998). An application of economic choice experiments to a traditional land use – Deer hunting and landscape change in the Scottish highlands. *Journal of Environmental Management, 52*, 335–351.
Cameron, T. A. (1988). A new paradigm for valuing non-market goods using referendum data: Maximum likelihood estimation by censored logistic regression. *Journal of Environmental Economics and Management, 15*, 355–379.

Cameron, T. A., & James, M. D. (1987). Efficient estimation methods for 'closed-ended' contingent valuation surveys. *The Review of Economics and Statistics, 69*, 269–286.

Canada Privy Council Office. (2001). *An accord between the Government of Canada and the Voluntary Sector Canada.* Voluntary Sector Task Force, ISBN 0-662-66216-4. Retrieved from http://www.vsi-isbc.org/eng/relationship/pdf/training_presentation.pdf

Cho, S. H., Yen, S. T., Bowker, J. M., & Newman, D. H. (2008). Modeling willingness to pay for land conservation easements: Treatment of zero and protest bids and application and policy implications. *Journal of Agricultural and Applied Economics, 40*, 267–285.

Ciriacy-Wantrup, S. V. (1947). Capital returns from soil-conservation practices. *Journal Farm Economics, 29*, 1181–1196.

Clawson, M., & Knetsch, J. (1966). *Economics of outdoor recreation.* Baltimore: Johns Hopkins University Press.

Concu, G. B. (2005). *Are non-use values distance-independent? Identifying the market area using a choice modeling experiment* (No.WP6M05). Retrieved December 2005, from http://www.uq.edu.au/rsmg/WP/WPM05_6.pdf

Concu, G. B. (2006). *Investigating distance effects on environmental values: A choice modeling approach.* Contributed paper prepared for presentation at the International Association of Agricultural Economists Conference, Gold Coast, Australia.

Correll, M. R., Lillydahl, J. H., & Singell, L. D. (1978). The effects of green belts on residential property values: Some findings on the political economy of open space. *Land Economics, 54*, 207–217.

Coursey, D. L., Hovis, J. L., & Schulze, W. D. (1987). The disparity between willingness to accept and willingness to pay measures of value. *Quarterly Journal of Economics, 102*, 679–690.

Darling, A. H. (1973). Measuring benefits generated by urban water parks. *Land Economics, XLIX*, 22–34.

Davis, R. K. (1963). Recreation planning as an economic problem. *Natural Resources Journal, 3*, 239–249.

Davis, A. (1984). Property rights and access management in the small boat fishery: A case study from southwest Nova Scotia. In C. Lamson & A. J. Hanson (Eds.), *Atlantic fisheries and coastal communities: Fisheries decision-making case studies* (pp. 133–164). Halifax: Dalhousie Ocean Studies Programme.

DeMartino, G. (2000). *Global economy, global justice: Theoretical and policy alternatives to neoliberalism (Routledge studies in contemporary political economy).* New York: Routledge.

Desvousges, W. H., Johnson, F. R., Dunford, R. W., Boyle, K. J., Hudson, S. P., & Wilson, N. (1993). Measuring natural resource damages with contingent valuation: Tests of validity and reliability. In J. A. Hausman (Ed.), *Contingent valuation: A critical assessment* (pp. 91–159). Amsterdam: North Holland.

Dupuit, A. J. É. J. (1844). De la mesure de l'utilité des travaux publics, Annales des ponts et chausses. Second series 8. (R. H. Barback, Trans.). (1952). On the measurement of the utility of public works. *International Economic Papers, 2*, 83–110. Reprinted in Arrow, K. J., & Scitovsky, T. (Eds.). (1969). *Readings in welfare economics* (pp. 255–283). Homewood: Richard D. Irwin.

EFTEC. (2001). *Study on the valuation and restoration of damage to natural resources for the purpose of environmental liability.* Brüssel: European Commission, Directorate General Environment.

Ervin, D., & Mill, J. (1985). Agricultural land markets and soil erosion: Policy relevance and conceptual issues. *American Journal of Agricultural Economics, 67*, 938–942.

Forrest, D., Glen, J., & Ward, R. (1996). The impact of a light rail system on the structure of house prices: A hedonic longitudinal study. *Journal of Transport Economics and Policy, 30*, 15–29.

Forrest, D., Grime, K., & Woods, R. (2000). Is it worth subsidizing regional repertory theatre? *Oxford Economic Papers, 52*, 381–397.

Foster, V., & Mourato, S. (1999). *Elicitation format and part-whole bias: Do contingent valuation and contingent ranking give the same result?* (CSERGE Working Paper, GEC 99-17). Retrieved from http://www.uea.ac.uk/env/cserge/pub/wp/gec/gec_1999_17.htm

Frech, H. E., III, & Lafferty, R. N. (1984). The effect of the California coastal commission on housing prices. *Journal of Urban Economics, 16*, 105–123.
Freeman, A. M. (1979). The hedonic price approach to measuring demand for neighborhood characteristics. In D. Segal (Ed.), *The economics of neighborhood*. New York: Academic.
Garrod, G., & Willis, K. (1992). Valuing goods' characteristics: An application of the hedonic price method to environmental attributes. *Journal of Environmental Management, 34*, 59–76.
Garrod, G., & Willis, K. G. (1999). *Economic valuation of the environment: Methods and case studies*. Cheltenham: Edward Elgar Publishing.
Georgiou, S., Bateman, I., Cole, M., & Hadley, D. (2000). *Contingent ranking and valuation of water quality improvements* (CSERGE Discussion Paper 2000-18). University of East Anglia. Retrieved from http://www.uea.ac.uk/env/cserge/pub/wp/gec/gec_2000_18.htm
Green, P. E., & Srinivasan, V. (1990). Conjoint analysis in marketing research: New developments and directions. *Journal of Marketing, 54*, 3–19.
Greenley, D. A., Walsh, R. G., & Young, R. A. (1981). Option value: Empirical evidence from a case study of recreation and water quality. *Quarterly Journal of Economics, 96*, 657–673.
Hageman, R. (1985). *Valuing marine mammal populations: Benefits valuations in a multispecies ecosystem* (Technical Report Administrative Report LJ-85-22). Silver Spring: National Marine Fisheries Service Southwest Fisheries Center.
Hai, W. B. (2009). *Yangtze finless porpoises reappear in Nanjing*. Retrieved February 15, 2009, from http://blog.sina.com.cn/s/blog_58e37f800100cexc.html
Hallowell, I. A. (1943). The nature and function of property as a social institution. *Journal of Legal and Political Sociology, 1*, 115–138.
Hanemann, W. M. (1991). Willingness to pay and willingness to accept: How much can they differ? *American Economic Review, 81*, 635–647.
Hanemann, W. M. (1999). The economic theory of WTP and WTA. In I. J. Bateman & K. G. Willis (Eds.), *Valuing environmental preferences: Theory and practice of the contingent valuation method in the US, EU and developing countries*. Oxford: Oxford University Press.
Hanley, N., & Spash, C. L. (1993). *Cost-benefit analysis and the environment*. Cheltenham: Edward Elgar Publishing Limited.
Hanley, N., Shogren, J. F., & White, B. (1997). *Environmental economics in theory and practice*. Oxford: Oxford University Press.
Hanley, N., Wright, R., & Adamowicz, W. (1998). Using choice experiments to value the environment. *Environmental and Resource Economics, 11*, 413–428.
Hanley, N., Wright, R. E., & Koop, G. (2000). *Modeling recreation demand using choice experiments: Climbing in Scotland* (Discussion Paper). Department of Economics, University of Glasgow.
Hanley, N., Shogren, J. F., & White, B. (2001). *Introduction to environmental economics*. Oxford: Oxford University Press.
Hanley, N., Schläpfer, F., & Spurgeon, J. (2003). Aggregating the benefits of environmental improvements: Distance-decay functions for use and non-use values. *Journal of Environmental Management, 68*, 297–304.
Hannon, B. (1994). Sense of place: Geographic discounting by people, animals and plants. *Ecological Economics, 10*, 157–174.
Hardin, G. (1968). The tragedy of the commons. *Science, 162*, 1243–1248.
Harless, D. W. (1989). More laboratory evidence on the disparity between willingness to pay and compensation demanded. *Journal of Economics and Behavior Organization, 11*, 359–379.
Harrison, D., & Rubinfeld, D. L. (1978). Hedonic pricing prices and the demand for clean air. *Journal of Economics and Management, 5*, 81–102.
Hensher, D. A. (1994). Stated and preference analysis of travel choices: The state of practice. *Transportation, 21*, 107–133.
Hoehn, J. P., & Randall, A. (1989). Too many proposals pass the benefit cost test. *American Economic Review, 79*, 544–551.
IWC (International Whaling Commission). (2002b). *Report of the scientific committee*. International Whaling Commission, IWC/54/4.

Jiang, Y., Swallow, S. K., & McGonagle, M. P. (2005). Contest-sensitive benefit transfer using stated choice models: Specification and convergent validity for policy analysis. *Environmental and Resource Economics, 31*, 477–499.
Jim, C. Y., & Chen, W. Y. (2006). Recreation – Amenity use and contingent valuation of urban greenspaces in Guangzhou, China. *Landscape and Urban Planning, 75*, 81–96.
Jin, J., Wanga, Z., & Ran, S. (2006). Comparison of contingent valuation and choice experiment in solid waste management programs in Macao. *Ecological Economics, 57*, 430–441.
Jin, J. J., Wang, Z. S., & Liu, X. M. (2008). Valuing black-faced spoonbill conservation in Macao: A policy and contingent valuation study. *Ecological Economics, 68*, 328–335.
Jud, G. D., & Watts, J. M. (1981). Schools and housing values. *Land Economics, 57*, 459–470.
Keyserlingk, C., & Rhodes, B. (2007). Using contingent valuation in hypothetical settings: Estimating the WTP for an HIV/AIDS vaccine. *Journal of Interdisciplinary Economics, 18*, 71–89.
Kirshner, D., & Moore, D. (1989). The effect of San Francisco bay water quality on adjacent property values. *Journal of Environmental Management, 29*, 263–274.
Knetsch, J. L. (1990). Environmental policy implications of disparities between willingness to pay and compensation demanded measures of values. *Journal of Environment and Economics Management, 18*, 227–237.
Knetsch, J. L., & Sinden, J. A. (1984). Willingness to pay and compensation demanded: Experimental evidence of an unexpected disparity in measures of value. *Quarterly Journal of Economics, 99*, 507–521.
Kniivilä, M. (2006). Users and non-users of conservation areas: Are there differences in WTP, motives and the validity of responses in CVM surveys? *Ecological Economics, 59*, 530–539.
Krutilla, J. V. (1967). Conservation reconsidered. *American Economic Review, 57*, 787–796.
Langford, I. H., Skourtos, M. S., Kontogianni, A., Day, R. J., Georgiou, S., & Bateman, I. J. (2001). Use and non-use values for conserving endangered species: The case of the Mediterranean monk seal. *Environment and Planning A, 33*, 2219–2233.
Ledyard, J. O. (1987). Market failure. In J. Eatwell, M. Milgate, & P. Newman (Eds.), *The new Palgrave: A dictionary of economics* (Vol. 3). London: Macmillan.
Loomis, J. (1994). Total economics values of increasing gray whale populations: Results from a contingent valuation survey of visitors and households. *Marine Resource Economics, 9*, 275–286.
Loomis, J. (1996). How large is the extent of the market for public goods: Evidence from a nationwide contingent valuation survey. *Applied Economics, 28*, 779–782.
Loomis, J., & Larson, D. (1994). Total economic values of increasing gray whale populations: Results from a contingent valuation survey of visitors and households. *Marine Resource Economics, 9*, 275–286.
Loomis, J., Kent, P., Strange, L., Fausch, K., & Covich, A. (2000). Measuring the total economic value of restoring ecosystem services in an impaired river basin: Results from a contingent valuation survey. *Ecological Economics, 33*, 103–117.
Loureiro, M. L., & Ojea, E. (2008). Valuing local endangered species: The role of intra-species substitutes. *Ecological Economics, 68*, 362–369.
Malkin, J., & Wildavsky, A. (1991). Why the traditional distinction between public and private goods should be abandoned. *Journal of Theoretical Politics, 3*, 355–378.
Marshall, A. (1890). *Principles of economics* (Vol. 1). London: Macmillan.
Martin, F. (1994). Determining the size of museum subsidies. *Journal of Cultural Economics, 18*, 255–270.
Mataria, A., Giacaman, R., Khatib, R., & Moatti, J. P. (2006). Impoverishment and Patients' 'Willingness' and 'Ability' to pay for improving the quality of health care in Palestine: An assessment using the contingent valuation method. *Health Policy, 75*, 312–328.
Mathieu, L. F., Langford, I. H., & Kenyon, W. (2003). Valuing marine parks in a developing country: A case study of the Seychelles. *Environment and Development Economics, 8*, 373–390.
Mazzanti, M. (2001). The role of economics in global management of whales: Re-forming or re-founding IWC? *Ecological Economics, 36*, 205–221.

Mazzotta, M. J., & Opaluch, J. J. (1995). Decision making when choices are complex: A test of Heiner's hypothesis. *Land Economics, 71*, 500–515.
McCay, B. J., & Acheson, J. M. (1987). *The question of the commons: The culture and ecology of communal resources*. Tucson: University of Arizona Press.
McConnell, C. R., & Brue, S. L. (2005). *Economics: Principles, problems, and policies* (16th ed.). New York: McGraw-Hill/Irwin.
Miranowski, J., & Hammes, B. (1984). Implicit prices of soil characteristics for farmland in Iowa. *American Journal of Agricultural Economics, 66*, 745–749.
Mitchell, R. C., & Carson, R. T. (1989). *Using surveys to value public goods: The contingent valuation method.* Baltimore: Johns Hopkins Press.
Morales, D. J. (1980). The contribution of trees to residential property value. *Journal of Arboriculture, 6*, 305–308.
Moran, D. (1999). Benefits transfer and low flow alleviation: What lessons for environmental valuation in the UK? *Journal of Environmental Planning and Management, 42*, 425–436.
Morrison, M., Bennett, J., & Blamey, R. (1999). Valuing improved wetlands quality using choice modeling. *Water Resource Research, 35*, 2805–2814.
Munasinghe, M. (1993). *Environmental economics and sustainable development.* Washington, DC: The World Bank.
National Research Council. (1986). *Proceedings of the conference on common property resource management.* Washington, DC: National Academy Press.
Nelson, J. P. (1980). Airports and property values: A survey of recent evidence. *Journal of Transport Economics and Policy, 14*, 37–52.
Nelson, J. P. (1982). Highway noise and property values: A survey of recent evidence. *Journal of Transport Economics and Policy, 16*, 117–138.
OECD. (1995). *The economic appraisal of environmental projects and policies: A practical guide.* Organization for Economic Co-operation and Development (OECD), Paris in co-operation with the Economic Development Institute (EDI) of the World Bank and Overseas Development Institute (ODI).
Ostrom, E. (1990). *Governing the commons: The evolution of institutions for collective action.* New York: Cambridge University Press.
Palmquist, R. B., & Danielson, L. E. (1989). A hedonic study of the effects of erosion control and drainage on farmland values. *American Journal of Agricultural Economics, 71*, 55–62.
Pate, J., & Loomis, J. (1997). The effect of distance on willingness to pay values: A case study of wetland and salmon in California. *Ecological Economics, 20*, 199–207.
Pavlova, M., Groot, W., & Merode, G. V. (2004). Willingness and ability of Bulgarian consumers to pay for improved public health care services. *Applied Economics, 36*, 1117–1130.
Pearce, D. W., & Markandya, A. (1989). *Environmental policy benefits: Monetary valuation*. Paris: Organization for Economic Co-operation and Development.
Plottu, E., & Plottu, B. (2007). The concept of total economic value of environment: A reconsideration within a hierarchical rationality. *Ecological Economics, 61*, 52–61.
Poor, P. J., & Smith, J. M. (2004). Travel cost analysis of a cultural heritage site: The case of historic St. Mary's city of Maryland. *Journal of Cultural Economics, 28*, 217–229.
Racevskis, L. A. (2005). *Public preferences and nonmarket values for the management of forest ecosystem services*. Doctoral Dissertation, Michigan State University.
Randall, A., Hoehn, J. P., & Tolley, G. S. (1981). *The structure of contingent markets: Some empirical results*. Paper presented at the Annual Meeting of the American Economic Association, Washington, DC.
Ready, R., & Navrud, S. (2002). Why value cultural heritage? In S. Navrud & R. Ready (Eds.), *Valuing cultural heritage: Applying environmental valuation techniques to historic buildings, monuments and artifacts* (pp. 10–28). Cheltenham: Edward Elgar.
Ridker, R. G., & Henning, J. A. (1967). The determinants of residential property values with special reference to air pollution. *The Review of Economics and Statistics, 49*, 246–257.
Roskill Commission. (1971). *Report of the commission on the third London airport*. London: HMSO.
Samuelson, P. A. (1947). *Foundations of economic analysis*. Cambridge: Harvard.

Samuelson, P. A. (1954). The pure theory of public expenditure. *The Review of Economics and Statistics, 36*, 387–389.

Samuelson, P. A. (1955). A diagrammatic exposition of a theory of public expenditure. *The Review of Economics and Statistics, 37*, 350–356.

Samuelson, P. A., & Nordhaus, W. D. (1989). *Economics: An introductory analysis*. New York: McGraw Hill.

Samuelson, P. A., & Nordhaus, W. D. (2005). *Economics*. New York: McGraw-Hill/Irwin.

Schafer, R. (1979). Racial discrimination in the Boston housing market. *Journal of Urban Economics, 6*, 176–196.

Schlager, E., & Ostrom, E. (1992). Property rights regimes and natural resources: A conceptual analysis. *Land Economics, 68*, 249–262.

Schulz, W., & Wicke, L. (1986). Die Kosten der Umweltverschmutzung – Beispiel: Waldsterben. *Umwelt und Energie, 7*, 52–85.

SDG. (1999). *Bus station passenger preferences*. Report by Steer Davies Gleave for London Buses, Transport for London, Windsor House, 42 – 50 Victoria Street, London SW1H 0TL.

SDG. (2000). *London underground customer priorities network*. Report by Steer Davies Gleave for London Underground, Windsor House, 42 – 50 Victoria Street, London SW1H 0TL.

Sidgwick, H. (1885). The scope and method of economic science. In H. Sidgwick (Eds.). (1904). *Miscellaneous essays and addresses* (pp. 170–199). London: Macmillan.

Sidgwick, H. (1901). *The principles of political economy* (3rd ed.). London: Macmillan.

Silberberg, E., & Suen, W. (1978). *The structure of economics: A mathematical analysis*. London: McGraw Hill International.

Simpson, R. D., Sedjo, R. A., & Reid, J. W. (1996). Valuing biodiversity for use in pharmaceutical research. *Journal of Political Economy, 104*, 163–185.

Stevens, T. H., Echeverria, J., Glass, R. J., Hager, T., & More, T. A. (1991). Measuring the existence value of wildlife: What do CVM estimates really show? *Land Economics, 67*, 390–400.

Stiegler, S. E. (1976). *Dictionary of economics and business*. Cambridge: Cambridge University Press.

Sutherland, R. J., & Walsh, R. (1985). Effect of distance on the preservation value of water quality. *Land Economics, 61*, 281–291.

Turner, R. K. (1999). The place of economic value in environmental valuation. In I. J. Batemann (Ed.), *Valuing environmental preferences: Theory and practice of the contingent valuation method in the US, EU, and developing countries* (pp. 17–41). Oxford: Oxford University Press.

Turner, K. R., Pearce, D., & Bateman, I. (1994). *Environmental economics: An elementary introduction*. Hemel Hempstead: Harvester Wheatsheaf.

Tversky, A., & Shafir, E. (1992). Choice under conflict: The dynamics of deferred decision. *Psychological Science, 3*, 358–361.

United Nations Environment Programme (UNEP). (1995). *Global biodiversity assessment*. Cambridge: Cambridge University Press.

Varian, H. (1987). *Intermediate microeconomics: A modern approach*. New York: W.W. Norton.

Wabe, J. S. (1970). A study of house prices as a means of establishing the value of journey time, the rate of time preference and the valuation of some aspects of the environment of the London metropolitan region. *Applied Economics, 3*, 247–255.

Wang, H., & Mullahy, J. (2006). Willingness to pay for reducing fatal risk by improving air quality: A contingent valuation study in Chongqing, China. *Science of the Total Environment, 367*, 50–57.

Weisbrod, B. (1964). Collective-consumption services of individual-consumption goods. *Quarterly Journal of Economics, 78*, 471–477.

Willis, K. G., & Benson, J. F. (1989). Recreational values of forests. *Forestry, 62*, 93–110.

Wood, S., & Trice, A. (1958). Measurement of recreation benefits. *Land Economics, 34*, 195–207.

Xu, Z. M., Cheng, G., Zhang, Z., Su, Z., & Loomis, J. (2003). Applying contingent valuation in China to measure the total economic value of restoring ecosystem services in Ejina region. *Ecological Economics, 44*, 345–358.

Yang, W., Chang, J., Xu, B., Peng, C., & Ge, Y. (2008). Ecosystem service value assessment for constructed wetlands: A case study in Hangzhou, China. *Ecological Economics, 68*, 116–125.

Chapter 4
Questionnaire Development

Abstract Based on the data obtained from two focus groups, the structure and essential components of the CVM questionnaire are developed in this chapter. Furthermore, various alternative approaches to elicit WTP questions are explored and one proper approach is selected, along with follow-up questions to help establish the validity of WTP responses. Before implementing the questionnaire survey on a large scale, a pilot survey was carried out to pretest the questionnaire survey. According to the feedback from the pretests, some modifications were made to the questionnaire (Sect. 4.3). In addition, sampling frame and sample size is presented (Sect. 4.4). All the possible survey methods are compared and one group-based survey method was selected (Sect. 4.5).

Keywords Focus group • Questionnaire • Validation • Sample frame • Survey method

4.1 Focus Group

Within this study, focus groups were organized in order to pursue the following goals: they helped to pretest the information folder and hypothetical market scenario; elicit the participants' knowledge about the YFPs; discuss the participants' understanding of and attitudes toward porpoise conservation; indicate whether they would contribute to it—to define the participants' favorite payment method and how much they might be willing to pay for it.

Before I describe the implementation of the focus group in our case study of China, some background information about this method will be given.

4.1.1 Introduction to Focus Group

Focus groups are frequently used to better understand how people consider an issue. Participants are selected because they have one common characteristic that is relevant with the topic of a focus group (Krueger and Casey 2000).

Y. Dong, *Contingent Valuation of Yangtze Finless Porpoises in Poyang Lake, China*,
DOI 10.1007/978-94-007-2765-6_4, © Springer Science+Business Media Dordrecht 2013

Focus group work was not systematically developed as a research technique within the social sciences until 1990s. Kitzinger (1994) gave the first systematic definition of focus groups as follows:

> Focus groups are group discussions organized to explore a specific set of issues .... The group is 'focused' in the sense that it involves some kind of collective activity.

Kitzinger (1995) sums up the essential role of the focus group technique as follows:

> The idea behind the focus group method is that group processes can help people to explore and clarify their views in ways that would be less accessible in a one to one interview ... when group dynamics work well the participants work alongside the researcher, taking the research in new and often unexpected directions.

The purpose of the focus group is to promote self-disclosure so that group participants' individual and unique understanding of an experience can be obtained.

In CV surveys, focus groups are designed to provide insights into respondents' likely understanding of and attitude toward the environmental goods or issues being investigated. Some use them to guide questionnaire design. Participants, around 10–15 in. each focus group, elicit their knowledge about the theme, discuss their understanding of the context of the environmental good, the good itself, its "values," who should provide it, how it should be paid for, whether they would contribute, and how much they might be willing to pay. Focus group sessions typically last between 1 and 2 h, and can provide valuable information in framing and designing a CV questionnaire survey (Garrod and Willis 1999).

4.1.2 *Implementation of Focus Groups in China*

The purpose of focus groups in this study was to test the information folder, to help guide the design of the information folder/questionnaire, to determine the payment vehicle, to define the payment range, and to establish the proper elicitation format of the WTP. Two focus groups were organized. One took place on 25 March 2009 in Nanbeigang village, which lies in the Poyang Lake region of China. Another took place on 1 April 2009 in Shentangyu village, situated in Beijing (see Fig. 4.1).

In total, 30 participants attended the focus groups, recruited from the two villages with the help of the village committee leaders. The demographic characteristics of participants in focus groups are summarized (see Appendix A). The group discussion lasted about 2 h. The process of the focus group meetings was organized according to the following procedure (see Box 4.1).

Fig. 4.1 Sites for focus groups in China (Source: http://www.topchinatrip.com/images/China_Map/China/China_Map.jpg)

Box 4.1 Focus Group Schedule

Step 1: About 10 min
Introduction presentation (personal information, brief introduction of UFZ, aim of focus group.)

Step 2: About 10 min
Participants fill in demographic form (see Appendix D).

Step 3: About 20 min
Question: Have you heard of the YFPs?
To go through the basic background information (see Appendix D).

Question: Do you understand the information folder? Do you still have any question? What information is most surprising to you?

Step 4: About 20 min
Question: Do you think they are important species? Are they worthy of being well-protected? Why? Any other reason?

(continued)

Box 4.1 (continued)

To look through threats faced by the YFPs and why to protect them (see Appendix D).
Question: What do you think about these threats? Do you think something should be done to protect the YFPs?

Step 5: About 40 min
Introduction to possible future scenario for the YFPs (see Appendix D)
To explain the future protection project of the YFPs (see Appendix D) and let participants answer the following questions.

Question:
Do you think it is feasible to establish the national nature reserve for the YFPs in Poyang Lake? If NO, do you have any alternative method?
Do you have any questions? Is anything unclear?
You can put forward any comments.

Step 6: About 20 min
To answer several questions:
How should it be paid for in your views? Donation, water bills?
Do you pay income tax? What is the tax rate? Except for income tax, do you pay for any other tax?
Do you pay water bills? If so, how much do you need to pay for water bills per month on average?
How often do you donate per year? How much have you donated most at one time at most?
Do you have some knowledge of the trust fund? Have you invested in it before?

Overall, participants were intensely involved in the discussion. Participants in Nanbeigang village, located in the Poyang Lake region, were more active than those in Shentangyu village, located in the Beijing region, which was positively related with their knowledge of the YFPs. The focus groups were fruitful and lots of information was collected. In the next section, the results of the focus groups will be explained and analyzed.

4.1.3 Analysis and Interpretation of Focus Group Findings

The discussion in the focus groups concentrated on several topics, which cover all the questions and items that were brought up by the participants. These topics include people's knowledge and perception of YFPs, people's views toward to protection of YFPs and their reaction to threats of the YFPs caused by human activities, and the acceptance of the hypothetical market. Concerning the acceptance of the hypothetical market, three smaller topics were discussed, including the establishment of a nature reserve, the payment method, and WTP and its potential problems.

It was planned to recruit the participants from the countryside in order to satisfy the representativeness of the population. In the end, all participants were farmers. On the whole, individuals found the YFPs were important and thought they were worth being protected for different reasons. For example, a number of respondents supported the protection of porpoises on the ground that the YFPs were the Second Category of National Key Protected Wildlife species in China. Other respondents living near the Poyang Lake took it as a great honor to live close to the YFPs. Compared with respondents in the village of Beijing, participants in the village of Poyang Lake embraced more information on the YFPs and could put forward some detailed suggestions on protecting the YFPs.

When discussing the great population decrease incurred by threats to the YFPs, participants in the village of Beijing showed some sympathy mixed with innocence while participants in the village of the Poyang Lake region insisted on water pollution being the main factor and denied that illegal fishing and overfishing could be a reason. Water pollution is closely related with fishery, which has direct influence on local residents' living. All participants expressed that they supported the protection of YFPs. Meanwhile, many of them insisted that the government should take the main responsibility of protecting the YFPs.

The idea of establishing a nature reserve received support from most participants both in Shentangyu (Beijing) and Nanbeigang (Poyang Lake region). Several individuals in Nanbeigang also indicated hesitation and were worried about the proposal. For instance, they were interested in whether the establishment of a nature reserve could help relieve water pollution, such as removing the chemical factories and banning waste water discharge. They were also worried about the problem that establishing a nature reserve would reduce their fishing area. If that happened, they asked whether they would receive some compensation.

In the focus group, income tax was proposed as the payment method. The majority of respondents did not like income tax as a payment vehicle based on two arguments: on the one hand, they were all farmers and were not used to monetary payments; on the other hand, there is a new rural tax-free reform. Therefore, donation was suggested as a proper payment method, and it was taken into account in the questionnaire design.

Only a small proportion of respondents told their WTP bids. The WTP response rate was extremely low, which should be regarded closely. There are several possible reasons for this phenomenon. First, a very unpopular payment vehicle, namely the income tax, could be responsible for the low response rate. Secondly, people are used to thinking that the protection of rare species is the government's task. Thirdly, the annual payment should be replaced by a one-off payment because of uncertainties on individuals' income. Finally, people are not familiar with such protection projects and WTP questions. In a future survey, all these factors should be considered when designing the questionnaire and selecting the survey method. (See Appendix B for detailed focus group summaries.)

4.1.4 Interviews with Experts

In addition to the focus groups, some porpoise experts in Wuhan,[1] governmental officials and Poyang Lake researchers in Nanchang[2] of China were interviewed in order to develop suggestions on the protection of YFPs that can support designing the hypothetical market. The interviewed government officials are familiar with managing nature reserves. The list of experts can be found in Appendix C. Basically, all experts showed positive attitudes toward the protection of YFPs. They agreed that the YFPs live in an urgent situation and need immediate conservation. They also indicated that they definitely support the establishment of a nature reserve, a most reasonable conservation measure at present and they could accept the hypothetical market scenario.

The interviews included the future population change of the YFPs, experts' perceived importance of the YFPs, overviews on threats to the YFPs, current and suggested conservation measures on the YFPs, and the question of whether Poyang Lake is the best habitat for the YFPs. Experts agreed that the porpoise population has diminished significantly and is still decreasing at a tremendous rate. In their view, the YFPs hold the same ecological functions as giant pandas but receive much less care than them. Since river dolphins have become extinct, the YFPs should be given more attention to, as they are the best substitutes of river dolphins. Poyang Lake is the best habit for the YFPs based on the relatively good water quality and large population of the YFPs. All these points were included in the information folder, helping individuals to gain knowledge of the YFPs.

Some threats such as shipping traffic were confirmed and potential additional threats were mentioned, such as the development of a water project in Poyang Lake, which will be used as a typical example of threat in the hypothetical market. The location and area of a possible nature reserve should be adjusted according to the experts' suggestions, which acted as an important reference in the hypothetical market development.

In addition to direct protection measures, indirect protection measures were also emphasized. The expert made it clear that fishermen's social welfare should be improved and guaranteed in order to completely prevent overfishing and illegal fishing. Economic compensation should be paid to fishermen when limiting their fishing. This should also be explained in the hypothetical market so as to make it as acceptable and realistic as possible.

---

[1] Wuhan, the capital city of Hubei Province, is a city in central China. It lies at the intersection of the middle reaches of the Yangtze River and Han River. The Institute of Hydrobiology (IHB), Chinese Academy of Sciences (CAS), the most authoritative institute in YFP research, is located in the city.

[2] Nanchang, the capital city of Jiangxi Province, is bordered by Poyang Lake on the east side. Jiangxi Fisheries Administration Office, which is responsible for the management of Poyang Lake and Jiangxi Normal University, where there is professional research group on Poyang Lake, are located in the city.

4.1.5 *Summary*

The focus groups and expert interviews led to several important conclusions: The proposal of establishing a nature reserve needed refining. The payment vehicle should be a donation or a trust fund instead of the income tax; WTP should be changed into a one-off payment because people have insecurity regarding their future income. All participants were farmers, which did not meet the representativeness of the sample population. This requires that the survey area should be adjusted in the final questionnaire survey. Based on the information gained from the focus groups and the expert interviews, I proceed to the questionnaire development in detail in the following section.

4.2 Questionnaire Structure

The questionnaire developed in this study consists of four parts. It includes: (1) general questions on environmental issues in order to test respondents' environmental attitudes, (2) questions on socioeconomic characteristics in order to better understand respondents' answers, (3) an information folder providing respondents with information about the YFPs, and (4) a hypothetical market in which non-market environmental goods and services are made available to respondents. The general environmental questions and the socioeconomic questions can be found in Appendix D. These questions can be used to examine whether the expected relationships between WTP and influencing variables hold. In the following, the main task is to explore the development of the information folder and the hypothetical market.

4.2.1 *Information Folder*

The information folder contains basic information about the YFPs, from physical features to protection status. It is a crucial component of the CV survey because it informs the respondents of what they are purchasing (see Appendix D).

Respondents' understanding of the environmental issue is at stake for the CV survey. According to psychology research, individuals only consider factors that are explicitly conveyed in the information sets and tend to ignore information that needs to be logically inferred (Harris et al. 1989). Furthermore, respondents are prone to concentrating on preconceived notions or certain aspects of the provided information to deal with the decision task. Clearly delineating the exact character of the products of environmental management is a prerequisite for any valuation task, especially where the constructs of concern are the less-tangible outputs of wildlife management, scenic/visual resources, wildland recreation, and so on (Harris et al. 1989). The validity and reliability of WTP depend on the amount and type of information individuals hold. The more detailed the information, both acquired through personal experience and provided by the survey, the more valid and reliable the WTP statements made by respondents are (Whitehead et al. 1995). Cummings et al. (1986) specify that the accuracy of WTP statements increases with the respondent's familiarity with the natural resource. Tisdell and Wilson (2006) confirm that WTP depends not only on how well-known the species are but also on how clear the provided information about the species is. Subsequent changes in individuals' knowledge can alter the willingness to contribute funds for species' protection.

Since individuals are likely to be very unfamiliar with the YFPs, considerable efforts have to be made to design a detailed information folder to increase their acquaintance with them. It is essential to offer a simple and broad information folder to respondents, which helps respondents understand the environmental issue. Since the information folder plays an essential role, if people have little knowledge of the issue or have never even heard of it, it is necessary to describe the environmental good of the porpoise protection comprehensively and precisely. The aim is to make respondents comprehend the environmental issue as easily and quickly as possible so that they can offer their true WTP statements for the porpoise conservation project.

The information folder developed for the YFPs consists of six parts. The first three parts are basic background information, including the physical description, distribution, and protection status of the YFPs. Physical description is vital for the respondents, who have never heard of or seen YFPs. The description includes some information on size, weight, and skin color, and one accessory picture. People should be able to distinguish the YFPs from river dolphins. River dolphins are more famous than the YFPs and they have similar appearance. Therefore, some people may mistake YFPs for river dolphins. Distributed information tells respondents where the YFPs can be found in the world. Protection status information provides the population change and density change both on the whole and in different branch habitats, which notify respondents that the porpoise population is declining at a tremendous rate and that this species needs urgent protection.

The fourth part deals with threats faced by the YFPs, which facilitate respondents in understanding why the porpoise population is decreasing dramatically. In this part, descriptions of threats are combined with caricatures. Some figures are taken from some news reports, such photos of dead YFPs and a schematic picture of the Poyang Lake project. The case of river dolphins is cited in order to make people sensitive to the fact that the YFPs would meet the same fate as river dolphins if no effective conservation measures are taken immediately.

The fifth part is to tell respondents why to protect the YFPs. In this part, special characteristics with the YFPs are listed and explained. This is crucial because it is decisive for respondents to give their right judgment on porpoises' use and non-use values. The specialty and functions of the YFPs are highlighted here.

The sixth part is to explain the situation of the YFPs in Poyang Lake. In this study, Poyang Lake is locked up as the research field. As explained in the given information, the YFPs are distributed in the middle and lower reach of the Yangtze River system. Respondents may feel confused why the project merely concentrates on porpoise protection in Poyang Lake. Several arguments, gained from expert interviews, are put forward here in order to counteract respondents' confusion.

The seventh part is to forecast the possible development of the YFPs in two scenarios. One scenario is to leave them alone—then they would die out in 25–100 year or even a shorter time. A second scenario is to take effective protection measures right now, such as establishing a nature reserve. In this scenario, porpoise population would increase and then the risk of extinction could be avoided. At the same time, the lake ecology could be improved. Pictures are an important assistance in this part. Detailed information can be found in Appendix D.

4.2.2 Designing a Hypothetical Market

Generally, a hypothetical market is used when values are not directly visible in actual markets, as is the case with non-use values. The TEV of the YFPs cannot be directly observed in the market. To estimate these values, a hypothetical market is needed where people are asked to state their preference on porpoise conservation.

In this study, a realistic hypothetical market context is considered to be one that is based on the plan that a nature reserve for the YFPs will be built up in Poyang Lake. Overfishing and illegal fishing would be forbidden in this nature reserve area. Additionally, sand mining would be limited in the water area outside the nature reserve area. Fisherman could still fish regularly. A sustainable economic development mode is recommended. Against this background, the following scenario was developed:

> In order to protect the YFPs efficiently, a special national nature reserve should be built up for the YFPs in Poyang Lake. The nature reserve would cover n area of about 300 km^2. Nature reserve 300 km2. In the nature reserve, there would be no risk of entanglement with fishing gear or crashes with sand-mining machines so that porpoises can navigate freely and safely. That is, the habitat conditions for the YFPs would come to a good state so that the porpoises can reproduce smoothly. Small habitats would not be suitable for the porpoises. We would transfer the porpoises living in the north of Duchang Dam to the nature reserve. It is estimated that the porpoise population would increase from 400 to 600 in the next 10 years in the nature reserve. This would mean that the YFPs would not be at risk of extinction any more.

In the scenario, the establishment of the nature reserve would be authorized by the Ministry of Agriculture of the People's Republic of China.

4.2.3 Obtaining WTP

Because porpoise conservation is a public good, many respondents do not have experience in deciding a price on a public good. Therefore, a proper payment method should be proposed and an elicitation format should be offered so as to make respondents comfortable with giving a true statement.

4.2.3.1 Payment Vehicle

In CVM questionnaires, a "payment vehicle" is usually specified as a means of securing an environmental outcome (Cummings et al. 1986). The payment vehicle provides the context for payment. It affects the way respondents answer the WTP question (Morrison et al. 2000). An inappropriate choice of the payment vehicle may lead to, or increase biases. When a respondent misperceives a payment vehicle, he/she may give an untrue statement and then a bias arises (Mitchell and Carson 1989). Therefore, to choose a proper payment vehicle is very important for obtaining realistic WTP values.

Payment vehicles are classified into voluntary payments and coercive payments. Income tax, value added or sales tax, trust funds, entry charges, property taxes, water bills, and utility bills belong to coercive payment vehicles (Garrod and Willis 1999). Donations and gifts belong to voluntary payment vehicles. Not all vehicles are viable options in a given situation. The choice of the payment vehicle should be governed by three considerations: (1) the suitability of the payment to the elicitation of use and/or non-use values; (2) the credibility of the payment vehicle; and (3) the acceptability of the payment vehicle (Green and Tunstall 1999). The chosen bid vehicle should have a plausible connection with the valued good, and should also be recognized to be fair and equitable. At the same time, the payment vehicle should be credible for respondents in order to minimize potential biases and protest responses. In addition, the payment vehicle should be acceptable in the given institutional and cultural context. An undesirable payment vehicle may strongly hinder respondents' estimates and increase protest responses.

According to the results of focus groups, voluntary donation was selected as the payment vehicle in order to elicit WTP for the protection of the YFPs. During the focus group discussions, participants recommended that voluntary donation or a trust fund should be used as the payment vehicle. Some people suffer psychological burden from mandatory payment, such as taxes or entrances fees. Individuals complain about taxes because they are often asked to pay all kinds of taxes, and they think many taxes are implausible. Voluntary donations can help them release such psychological burden of being forced to pay. Individuals think that through donation they can decide their WTP bids completely according to their own preference. Champ et al. (1997) argue that donations are more useful payment vehicles for CVM, because they offer a plausible means of estimating the economic value of small-scale public goods, while survey respondents may object to mandatory payment schemes such as taxes or entrance fees. Donation vehicles have also been applied in several CV studies (Duffield and Patterson 1992; Navrud 1992; Seip and Strand 1992; Swallow and Woudyalew 1994; Bateman et al. 1993; Loomis and Gonzalez-Caban 1997; Spencer et al. 1998). In China, people have many experiences with voluntary donations to Project Hope[3] and to people who suffer from natural emergencies and disasters, such as earthquakes. Therefore, donation is a proper payment vehicle in the CV study of the YFP conservation.

---

[3] *Project Hope* is a Chinese public service project organized by the China Youth Development Foundation (CYDF) and the Communist Youth League (CYL) Central Committee. Started on October 30, 1989, it aims to bring schools into poverty-stricken rural areas of China, to help children whose families are too poor to afford complete elementary school education. Through Project Hope, the CYDF has also sought to improve educational facilities and improve teaching quality in poorer regions.

4.2.3.2 Individual or Household Payment

Respondents may be asked to pay in the unit of individual or household. Individual payments may be problematic when the respondent is not economically independent. Household payments may pose problems when the respondent is uncertain about the household budget and about other household members' values (Bateman et al. 2002a). So only the head of the household is suitable for answering the household payment but it cannot be guaranteed that all respondents are the heads of households. Some CVM studies apply individual payments (Loomis and Larson 1994; Wang and Mullahy 2006; Mathieu et al. 2003; Boxall et al. 1996; Jim and Chen 2006). Individual or household payment also depends on the choice of payment vehicles. For example, water bills are normally paid on the household level while an entry charge is levied on an individual level (Bateman et al. 2002a). Donation could be elicited either on the individual or the household level.

In China, there is still another point that should be taken into consideration when making the choice between individual payments and household payments. The definition of household in law is not consistent with that in actual life. In the Chinese language, the word "household" and the word "family" have the same translation—there is no clear borderline between them. In the focus group, participants were asked: "How many members are there in your household?" Many of them indicated that it was a confusing question. There is a special household registration record called *Hukou*. A *Hukou* refers to the system of residency permits, where household registration is required by law in the People's Republic of China. It is very common that one person works in a different city from where his/her *Hukou* is located. It is possible that family members have different *Hukou* but they are one family in actual life. They live together and the household budget is shared. It is also possible that people live in different cities with the same *Hukou* registration and their budget is separate. In such situations, household payments will be more difficult for respondents than individual payments. Simplifying the questions as much as possible is one important principle that designing the questionnaire should follow. Individual payment is, therefore, preferable and chosen in this CV study.

4.2.3.3 Timing of Payment

According to timing, payments can be divided into several types, such as one-off payment, annual payment, monthly payment, and even payment per visit. One-off payment tends to be appropriate for things that are largely provided by a one-time capital expenditure, such as the purchase of a wilderness area, the restoration of a cultural site, while annual payments have a propensity for goods like clean air and water that are provided continually. One advantage of one-off payment is that many of the sequencing problems can be eliminated because any new good that was once paid for does not enter the sequence of possible new goods (Bateman et al. 2002a). In addition, one-off payment resembles a donation framework while donation is selected as the payment vehicle in this CVM study. Besides, annual payment was originally used in the hypothetical market of porpoise conservation, but the participants in focus groups suggested that it was not plausible because they did not know their future income and they could not decide their future WTP according to their present income level. Hence, one-off payment is applied in the questionnaire.

4.2.3.4 Elicitation Format

Elicitation format is the way in which individuals are asked to state their maximum WTP for the environmental good, like the porpoise protection in this case. The main elicitation formats are as follows:

First, *an open-ended elicitation format* is a straightforward way of disclosing values. In this elicitation format, no value is specified and individuals are asked a simple question on their maximum WTP for the valued good; for example,

> Suppose some funds would be raised for building up a nature reserve in Poyang Lake in order to protect the YFPs well. What is the maximum amount you would be willing to pay for the project to increase the porpoise population?

The approach is relatively straightforward and easy to implement. Obviously, this kind of questions is easy for respondents who have some experiences in purchasing similar goods but not for respondents without any experience. For example, if some respondents have participated in CV questionnaire surveys on other rare species, they can gain some experience and may know how to answer a WTP question properly. If not, respondents may have some difficulty with this kind of format. Moreover, most of the daily goods are sold at fixed prices in supermarkets and shopping malls and customers are seldom asked to state their maximum WTP. As a consequence, the open-ended format is prone to produce a large number of unreliable responses to the WTP questions (Bateman et al. 2002a). For this reason, the National Oceanic and Atmospheric Administration (NOAA) report by Arrow et al. (1993) contended that open-ended formats should not be used to elicit values for environmental goods, for which there is no market, such as the protection of YFPs. Most respondents in China are unfamiliar with the valuation of the YFP protection and only few of them have done similar valuation before. Thus, open-ended elicitation format is not suitable for this CVM study. It can be concluded that open-ended elicitation is not an optimal choice for valuing porpoise conservation.

Secondly, a *dichotomous choice elicitation format* became increasingly popular in the 1990s (Bateman et al. 2002a). This approach simplifies the difficult cognitive task by presenting respondents with specific bids. It is more similar to market transactions, where people are accustomed to deciding whether to buy a good at a specific price (Nocera et al. 2002). There are mainly two kinds of dichotomous choice formats, including single-bounded and double-bounded dichotomous.

In the *single-bounded elicitation format*, only one payment amount is presented at one time and the respondent is asked to show whether he/she agrees with the amount or not;

> Suppose some funds would be raised for building up a nature reserve in Poyang Lake in order to improve the protection of the YFPs. Would you be willing to pay 100 RMB for the

project to increase the porpoise population? (The price is varied randomly across the sample.)

YES or NO

Similarly, two payment bids are offered at one time in the *double-bounded dichotomous choice method.*

Suppose some funds would be raised for building up a nature reserve in Poyang Lake in order to protect the YFPs well. Would you be willing to pay 100 RMB for the project to increase the porpoise population? (The price is varied randomly across the sample.)
If YES: Would you pay 200 RMB?
If NO: Would you pay 50 RMB?

Except for simplicity, this elicitation format also provides incentives for the truthful revelation of preferences under certain circumstances. The dichotomous choice elicitation format is less likely to yield strategic responses than open-ended elicitation format because of incentive compatibility[4] (Bateman et al. 2001). If the given price is not bigger than the respondent's WTP, he/she would accept the bid. Otherwise, the respondent would reject the bid.

However, the format has disadvantages, such as increased analytical demands on sensitivity analysis incurred by the variability of benefit measures (Kerr 1996). One of the most important concerns is that for a given level of estimation precision, DC studies require quite larger sample sizes than open-ended approaches because they collect less information from each respondent. DC survey elicits information only on whether a respondent's WTP lies above or below a nominated money amount (Bateman et al. 2001). A sample survey at a very large scale is not practical from the perspectives of time and budget in the YFP case. Consequently, DC elicitation is not recommended in this survey.

Thirdly, an *iterative bidding format* employs a single starting bid and follow-up discrete choice questions in which the starting bid is varied in small amounts until an equilibrium bid is attained (Whitehead 2002).

Suppose some funds need raising for building up a nature reserve in Poyang Lake in order to protect the YFPs well. Every person would be required to pay 100 RMB per year for the project to increase the porpoise population. Would you be willing to pay this fee?

YES or NO

If YES, Would you like to pay 110 RMB?	If NO, would you like to pay 90 RMB?
If YES, Would you like to pay 120 RMB?	If NO, would you like to pay 80 RMB?
...	...

Iterative bidding was one of the most widely used formats in the 1970s and 1980s. After that, this format was abandoned with the detection of start-point bias (Whitehead 2002). For example, Desvousges et al. (1987) implemented the iterative bidding format using different starting amounts. They found differences in final WTP amounts, and expressed caution against the further use of the iterative format. Therefore, iterative games are not recommended for assessing porpoise conservation.

[4] An allocative mechanism or institution is said to be incentive compatible when its rules provide individuals with incentives to truthfully and fully reveal their preferences (Cummings et al. 1997).

Fourthly, the *payment card* is a method in which respondents are presented with a visual aid card with a large array of potential WTP amounts. Respondents choose one of them that best represents their maximum WTP (Mitchell and Carson 1989). This method was developed by Mitchell and Carson (1981, 1984) on the basis of open-ended elicitation format and bidding games.

Suppose some funds are raised for building up a nature reserve in Poyang Lake in order to protect the YFPs well. What is the most you would be willing to pay for the project to increase the porpoise population (or to avoid the decrease of the porpoise population)? (Please circle one value)

1 RMB □	5 RMB □	10 RMB □	25 RMB □	30 RMB □
50 RMB □	60 RMB □	70 RMB □	80 RMB □	90 RMB □
100 RMB □	200 RMB □	300 RMB □	400 RMB □	500 RMB □
600 RMB □	700 RMB □	800 RMB □	900 RMB □	1000 RMB□

The advantage of a payment card is that the visual aid facilitates the valuation task by providing a bids context (Randall et al. 1981; Mitchell and Carson 1989). With this method, the starting point bias and outliers can be reduced in comparison to other elicitation formats (Bateman et al. 2002a). Yea-saying can be avoided because respondents can circle a rather low value instead of saying yes to a particularly denoted value. The payment card method was suspected to be vulnerable to biases associated with the ranges listed on the card, because the presented payment levels may not exactly represent the respondent's value (Mitchell and Carson 1989). However, Rowe et al. (1996) found that the payment card format can be seen as free of a range bias when the range of the WTP distribution included on the payment card is sufficiently large so that it does not constrain the respondent. This suggests that the range of values displayed on the card should be based on pretest surveys. It holds high credibility, because it leads to desirable response properties and to efficiencies in data collection (Reaves et al. 1999). So in this study, I decided to use the payment card method for valuing the protection of YFPs.

One issue that remains is the question how many bids a payment card should contain. Maintaining the range of a payment card and increasing the number of entries to reduce the interval size may be necessary for higher precision, though it seems clumsy for respondents. The payment card for the protection of YFPs contains 20 values. The payment amounts range from 1 to 1,000 RMB. The payment amounts are based on the findings of the focus groups (see Sect. 4.1).

Suppose some funds are raised for building up a nature reserve in Poyang Lake in order to protect the YFPs. There are 400 porpoises living in Poyang Lake and the population will increase to 600 in 10 years. What is the most you would be willing to pay for the project to increase the porpoise population?

______________RMB/Year

If you find it difficult to decide, you can refer to the following amounts (please circle one value):

1 RMB □	5 RMB □	10 RMB □	25 RMB □	
50 RMB □	60 RMB □	70 RMB □	80 RMB □	90 RMB □
100 RMB □	200 RMB □	300 RMB □	400 RMB □	500 RMB □
600 RMB □	700 RMB □	800 RMB □	900 RMB □	1000 RMB □

One direct question is placed before listed values. This is in line with the suggestions proposed by focus group participants. Some participants suggested that they felt pressured with the numbers on the card, and one direct question can help eradicate the pressure by providing freedom of deciding WTP amount independently. Although the direct question is difficult for many respondents, it is still hoped that respondents can decide the WTP bid on their own with the intention of avoiding the range bias and anchoring bias from the payment card.

Until now, the valuation scenario, payment vehicles, and elicitation formats have been explored and selected. The complete hypothetical market is developed and can be found in Appendix D. Also a complete questionnaire is accomplished now. Whether the questions in a CVM questionnaire are consequential or not is measured by whether respondents answer truthfully. This is called the validity of responses. In the subsequent section, the validation of WTP will be discussed.

4.2.4 *Validation of WTP*

The validity of a measure is the degree to which it measures the theoretical construct under investigation. This construct is, in nature of things, unobservable; all I can do is obtaining imperfect measures of that entity (Mitchell and Carson 1989). In the CV context, the theoretical construct is the maximum amount of money the respondents would actually pay for the public good, such as porpoise conservation in this study, if the appropriate market existed. The ultimate test of the accuracy of CVM is whether respondents will actually pay the amounts that they tell in a CV study.

In order to make sure that respondents consider the WTP question thoroughly and give a thoughtful answer, they are asked to explain why they are (not) willing to pay for the porpoise conservation project. After answering that they are willing or unwilling to pay for the project, they are asked to give the reasons for their willingness or unwillingness by choosing from the options listed on the questionnaire. Also before answering WTP question, respondents are reminded of their budget constraints and are also told that if not enough money is collected, the nature reserve will not be built up. This was done in order to encourage respondents to give the most accurate response. The follow-up debriefing question after WTP questions requires that respondents state how they decide the WTP amount. Their statements help identify invalid responses, such as protest responses, strategic bids, embedding effects, which will be explained in detail in Chap. 5.

4.2.4.1 Categories of Validation

The American Psychological Association (APA) (1974) classified validity into three types: content validity, criterion validity, and construct validity.

Content validity refers to whether the construct's [5] domain is completely covered. It involves both the appropriate framing of the study and questions asked in relation to the good being valued. It is difficult to judge because it depends on the subjective judgments of those examining the study, which is affected by people's knowledge and experience. Garrod and Willis (1999) listed the obvious criteria within content validity as follows:

- Is the description of the good being valued clear enough for respondents to comprehend well?
- Are the payment vehicle and elicitation format acceptable?
- Are there large numbers of protest bids?
- Have respondents been given enough time to think about answering the WTP question and explain the implications of their answer in the light of all other commitments on their budget and the availability of substitute goods to the one being valued?

The acceptability of the payment vehicle and the elicitation format were examined in the pretest in Beijing. Feedback from participants in workshops informed us whether the information folder was complete and sufficient and whether they had enough time to think about the questions included in the questionnaire.

Criterion validity is concerned with the comparison of CV estimates with actual market or simulated market experiments (Mitchell and Carson 1989). The hypothetical markets can produce hypothetical answers, which are not consistent with the answers produced in actual markets because there are lots of differences between hypothetical markets and actual markets. Many conditions in actual markets cannot be realized in hypothetical markets. For example, individuals are not penalized for "getting it wrong." It is impossible to prove the criterion validity of WTP estimates on porpoise conservation, since there is no actual market for the YFPs or any other similar rare species. Therefore, criterion validity could not be tested in our research.

Construct validity involves the degree to which the WTP measure relates to other measures predicted by theory. *Construct validity* is classified into *convergent validity* and *theoretical validity*. *Convergent validity* refers to the convergence between a CVM measure and travel-cost method (TCM) or hedonic price method (HPM) measure of the value of the same good (Garrod and Willis 1999). The test of convergent validity is only applied in the circumstances of measuring use values because TCM and HPM can only measure use values. In the case of porpoise conservation, non-use values are likely to be predominant. TCM and HPM cannot measure the TEV of the conservation of YFPs, which means that convergent validity could not be tested in this research.

[5] A *construct* in the philosophy of science is an *ideal* object (i.e., one the existence of which may be said to depend upon a subject's mind) as opposed to a *real* object (one the existence of which is not so dependent) (Bunge 1974). Here it refers to the hypothetical market built for one certain environmental good or service.

Theoretical validity involves assessing the degree to which the results of a CV study are consistent with theoretical expectations. For the purpose of testing the theoretical validity, the respondents' WTP bids are regressed against independent variables believed to be theoretical determinants of these values (Garrod and Willis 1999). Almost all valuation studies assess theoretical validity, which will also be examined in this CVM study.

4.2.4.2 Theoretical Validation

Theoretical validity seeks to answer the question whether the sign and size of the estimated coefficients are in accordance with theoretical expectations. This means that only those variables that are predicted by theory to determine an inverse demand function should be included in the regression equation, and precluding a data mining approach to obtain the highest R^2 value (Garrod and Willis 1999).

In statistics, the coefficient of determination, R^2, is used in the context of statistical models whose main purpose is the prediction of future outcomes on the basis of other related information. It is the proportion of variability in a data set that is accounted for by the statistical model (Steel and Torrie 1960). The coefficient provides a measure of how well future outcomes are likely to be predicted by the model. In linear regression, R^2 is simply the square of the sample correlation coefficient between the outcomes and their predicted values, or in the case of simple linear regression, between the outcome and the values being used for prediction. In such cases, the values range from 0 to 1 (Everitt 2002).

R^2 values can vary tremendously in regressing WTP or WTA against respondents' socioeconomic characteristics. In some studies, regression models only manage to explain 3 or 4% variance in WTP responses (Lienhoop and MacMillan 2007); while in others, 60% or more of the variance is explained (Bateman et al. 1995). Mitchell and Carson (1989) suggest that a CV study, which has an R^2 of less than 0.15, might be deemed unreliable. Low R^2 values are not necessarily evidence of theoretical inconsistency: They could merely be indicative of less tightly clustered observations around the regression line and stochastic variation in the data, which overshadows the systematic influence of variables. Low R^2 values are often associated with environmental goods that attract very small bids, given respondents' uncertainty and ambivalence in valuing non-market environmental goods. Since these goods are very often a minor part of respondents' real income, income itself does not explain variation in WTP: the consumption of such goods may depend purely upon income-independent preferences, which are more difficult to incorporate in a regression model, or consumption may even be purely stochastic (Garrod and Willis 1999). Nevertheless, high R^2 values do provide evidence of reliability.

Evidences of theoretical validity are highly desirable in CV when the valuation results serve for decision making or policy purpose. In any CV study, there are usually several expectations regarding the relationship between stated values and factors such as the uses of the good, the socioeconomic characteristics of the respondents, and the attitudes that they hold (Bateman et al. 2002a).

The relationship between the use of a good and stated values is expected to be positive and vary directly with the particular degree of use. In order to test the relationship, respondents are asked to answer whether they care about the uses of the YFPs before answering WTP question. Associations between stated values and reported attitudes, membership of interest groups, and other demographic indicators of concern regarding porpoise conservation can also be reasonably hypothesized. Many economists support the testing of such hypotheses as basic elements of theoretical validity analyses (Arrow et al. 1993; Desvousges et al. 1993; Schuman 1996).

In addition, attitudes regarding the equity, fairness, justice, and trust dimensions of the valuation scenario will impact upon stated values (Bateman et al. 2002a). Although these concerns constitute a bias in value estimates, prior expectations can still be formulated and tested. Details will be treated in Chap. 5.

4.3 Pretest and Modification of Questionnaire

Prior to the main survey, a pilot study was conducted to test the questionnaire and to assess whether the information and hypothetical markets were complete and understandable for participants to decide on their WTP bid. Specifically, it aims to test respondents' understanding of the information folder and the hypothetical market, to test whether the elicitation format and the payment vehicle are acceptable, and to test the bid distribution. All comments during the in-person interviews were recorded for further improvement of the questionnaire and the information folder. While the focus groups can be seen as elements to design the empirical study (see Sect. 4.1), the pretest serves the purpose to finally check the study design.

4.3.1 Characterization of the Pretest

The pretest of the questionnaire took place in Beijing from 23 to 25 September 2009. In total, 22 respondents were recruited in school yards and streets. Among the respondents, there are 10 males (45.5%) and 12 females (54.5%). Four of them are members in some environmental groups, such as WWF. Their age ranges from 16 to 65, and 20 of them belong to the age category of 16–35. The highest educational level that most respondents have is bachelor or above, except for one person that is illiterate and one person who only completed primary school. Their family size varies from 2 to 5 members. The income level ranges from category I (less than 500 RMB) to category X (over 10,000 RMB). On the whole, the sample covers a wide range and includes individuals with very different socioeconomic characteristics. Therefore, their overviews and comments on the questionnaire itself can mirror most individuals. If they can understand and finish the questionnaire, then the questionnaire can be applied in a survey on a large scale.

4.3.2 *Findings from the Pretest Survey*

During the pretest, individuals were confronted with the questionnaire and their comments were collected. In this subsection, (1) people's opinion and comments are explained in the order of general questions on environment and nature issues; (2) the information folder, and (3) the hypothetical market. At first, I will explain individuals' overall views on environment and nature.

4.3.2.1 Respondents' General Opinion on Environment and Nature

According to the pretest analysis, people showed significant concern about environmental issues. Fourteen respondents (63.6%) indicated that they watch/listen to programs about environment and nature whenever they can. Only two respondents hardly ever watch/listen to such programs. Ranking the policy priority on five issues, including (1) providing medical and endowment insurance in rural area, (2) lowering housing price, (3) solving the graduate unemployment, (4) education, and (5) environment and wildlife protectionenvironmental and wildlife protection was given the least priority. Solving the graduate unemployment issue was seen as the most urgent issue, according to the respondent's opinion. Lowering the housing prices was in the second position and education in the third. Providing medical and endowment insurance in rural areas was considered more urgent than environmental and wildlife protection—ranking in the fourth position. Of the respondents, 81.8% strongly agreed that environmental issues should be paid attention to during the process of economic development, and 9.1% of respondents agreed with that. Of the respondents, 77.3% of respondents strongly agreed and 13.6% agreed that people should invest in the environment for the benefits of future generations. Of the respondents, 45.5% of respondents thought that the endangered rare species should be protected by increasing their population, and 27.3% of respondents believed that the population should be protected by keeping their present population. Similarly, 63.6% of respondents strongly agreed and 27.3% of them agreed that animals have a right to live on Earth even if they provide no direct use for mankind.

As for the questionnaire's questions and answers, respondents could easily understand them. A little suggestion was made that internet should be included as the main media. Consequently, the question can be changed into:

> Q.1 Programmes about the environment and nature are often on TV, internet, and radio. Please indicate, which statement most accurately reflects your own opinion about these programmes (Cross one option only).

4.3.2.2 Information Folder on Yangtze Finless Porpoises

Generally speaking, respondents could grasp the general information of the YFPs. The majority of respondents feel that the provided information could make respondents understand that the YFPs were important species and are worth protecting. However, a few respondents hold a different opinion:

> "It is natural regulation. The species which do not adapt to the natural conditions will die out sooner or later. We can just leave them alone."
>
> "Even though the YFPs have been endangered, what can we do about them? In history, lots of species have died out. Take it easy."

At the same time, it seems that information cannot satisfy some respondents' knowledge need on the YFPs, and they suggested that more information should be supplemented, including biological features, their significance, protection measures, and how to resolve the conflict between economic development and porpoise conservation.

Regarding the biological features of the YFPs, some people suggested that more pictures of the YFPs should be added in the questionnaire due to the fact that pictures are simple and impressive. It was also recommended that a short movie should be made on the YFPs that could be played in every valuation workshop. Two individuals suggested that some more professional knowledge of the YFPs should be included, such as reproduction frequency, birth rate, life span, etc. However, these suggestions are not adopted in this case for several reasons: (1) Information overload may happen when more information is included; (2) there is no relevant research on reproduction frequency, birth rate, life span; (3) Only a few professional persons have such requirements while the research of our study aims at very general people.

Concerning the significance of the YFPs, a few respondents think that the listed values on the questionnaire are not completely convincing. They feel that the listed important features of the YFPs are slightly too theoretical but not very practical. They expect more information on the relationship between the YFPs and human beings. They would like to know what will happen to human beings if they lose the YFPs, except for their ecological function in biodiversity. The answer is that the loss of the YFPs would not bring obvious changes to human beings' life, just like the extinction of river dolphins. In addition, the practical significance of the YFPs to human beings, expressed by some respondents, can be understood as economic benefits to some extent while the economic benefits of YFPs are not presently existent. Therefore, these suggestions are not included in the questionnaire.

In relation to the protection of YFPs, people suggest that some protection measures should be illustrated in the information folder. Furthermore, they would like to know what they can do to support the protection of YFPs. These suggestions are useful. Since the protection of YFPs will be discussed in the following section of the hypothetical market, it is not included in the section of the information folder any more. Similarly, the conflict with economic development and the protection of YFPs is not repeated in the information folder either.

4.3.2.3 Hypothetical Market

1. Are people willing to support the conservation project?
   As for the proposal of establishing a nature reserve area, 18 respondents answered that they would support the conservation project based on four

kinds of reasons: First, they support the conservation project because they care about the YFPs directly. Some people indicated that they are in favor of the conservation project because the YFPs are important and may hold great potential unknown values. However, the YFPs are endangered. Therefore, it is necessary to take protection measures for them, while the project is the most feasible and efficient protection measure at this time. Secondly, some people suggest that the project should be advocated because human beings should live in harmony with nature. Animals, as an important part of nature, should be preserved. Sustainable development should be the final goal that people pursue and work on. Thirdly, some individuals approved the project on the basis of responsibility. They agreed that the decrease of porpoise population is caused by human activities. If economic conditions allow people should take responsibility and compensate on the YFPs. Finally, some individuals insist on public engagement being very important for the wildlife protection, such as the YFPs. Besides, one person's support on the project stems from the affection to his hometown situated in the Poyang Lake region.

Two respondents answered that they do not know whether to support the conservation project or not for the following reasons:

> "How will the project be carried out in detail? What is the final aim of the project? What is the possible environmental impact brought about by the project? Only 200 more porpoises in ten years is not a convincing argument. I need more detailed information about the project. Or else I cannot make my decision whether to support the project or not."
>
> "The inputs and outputs of the conservation project are not clear enough. I cannot decide whether to support it."

2. Are people willing to donate or not?
Sixteen respondents would like to contribute to the "YFP's fund" and four respondents refused to donate any money to the conservation project. Respondents would like to donate based on their preference on the uses of YFPs.
Six kinds of uses of YFPs are explained in the questionnaire, including (1) sightseeing, (2) ecological function, (3) option use, (4) altruistic use, (5) bequest, and (6) existence. Every kind of use received a preference from the individuals. Most individuals were concerned with the ecological function of maintaining biodiversity and the existence of the YFPs and were, therefore, willing to contribute to the project. A part of respondents also suggested their willingness to preserve the YFPs for their future generations. In conclusion, the uses of the YFPs are so attractive so that most individuals would like to contribute to their conservation by means of donation.
Nevertheless, four respondents refused to pay for the conservation project for some reasons. For instance, they were afraid that the money will not be used for the conservation project. A few people also suggested that they had no money. One person said that it was up to his personal values and his money would be spent on some other goods and services.
3. What are the WTP bids?
The donation amount ranges from 5 to 1,000 RMB. Only one person would like to donate 1,000 RMB and this response can be considered as valid because the

respondent can afford it. Therefore, there are 16 WTP results, and mean value of WTP is 120.94 RMB. Because 15 out of 16 responses were willing to pay no more than 100 RMB, it was decided to change the range of the payment card from 1 to 100 RMB in the final version of the questionnaire.

Typical explanations for their WTP bids are listed as follows:

> "There are still lots of other species needing protection. This is how much I can afford for the YFPs."
>
> "There are lots of donations. I am used to donating 100 RMB every time."
>
> "I do not know how the money will be used exactly. This is how much I would like to donate. To tell the truth, I would prefer to donate to Hope Project for the poor children in the same situation."
>
> "This is how much I would like to donate for the conservation project. I would like to know more about the organizations. I hope it is a convincing trust fund organization."

In summary, all the WTP bids were based on budget constraints, which proved to be an important factor responsible for determining WTP bids in many CVM studies. This was also found in Chap. 3.

A few respondents still provided some comments. For example, some respondents had not heard about WWF and the Wuhan Baiji Conservation Foundation before the survey. They were not sure whether these two organizations are dependable and trustworthy. They asked for more information about the two organizations to prove the dependability of the organizations. Then, one short information folder of these two organizations was prepared at hand. In the large-scale survey, it was shown and explained to the respondents who wanted to see it.

Summing up, no larger problems emerged with regard to the introductory and socioeconomic questions asked in the questionnaire and market contexts. In response to difficulties that emerged in the pilot survey, some small changes were made. The final version of the questionnaire can be found in Appendix D.

4.4 Implementation of the Questionnaire

The implementation of the questionnaire involves several topics, such as the determination of the target population, the establishment of the sampling frame, the sampling design, and the sample size.

4.4.1 Target Population

The target population is the entire set of individuals, to which findings of the survey are to be extrapolated. The target population consists of those who receive the benefits of the non-market impact in question. For geographically well-defined goods, such as the YFPs, the target population should include those who gain direct use values and those who hold non-use values. The people who enjoy direct use values should include local residents in the Poyang Lake region and visitors. Non-use values take the greater proportion of the TEV of the YFPs. Therefore, the population holding non-use values should be included more correspondingly. However, the determination of the population holding non-use values is not as straightforward as that of the population holding use values (Bateman et al. 2002a). Absolutely, the survey cannot include all the users due to the limitation of funds and time.

YFPs are the Second Order of Protected Key Animals in China, and they are globally unique. Therefore, the target population can be defined as the people worldwide. Generally speaking, the familiarity with the YFPs declines with the distance from them. The focus group results showed that local residents in Poyang Lake region are much more familiar with the YFPs than people in Beijing. However, it is still difficult to decide the "cut-off" for the target population and then shrink the target population. Given limited funds, it was decided to define the target populations to be the people in China. The survey was, therefore, planned in Beijing, Guangzhou, and Nanchang of China.

4.4.2 Sampling Frame

After determining the target population, the next step is to put together a list of the target population, known as the sample frame population, from which the sample will ultimately be drawn. Some examples of a sample frame population are: the dwelling units within a city, the voters registered in a city, the members of environmental groups. The sampling frame is required to be consistent with the target population. However, there is some trade-off between the cost and coverage of a sample frame: lists of specific, smaller populations may be more readily available than a list of the general population (Bateman et al. 2002a).

In addition, some other decisions should be made prior to the choice of the sampling frame, such as the survey method and the payment vehicle. If a telephone survey is conducted, the sample frame should be a telephone directory. If the payment vehicle are electricity bills, all the residential and commercial customers of an electricity utility could be the sampling frame. In the survey on the protection of YFPs, the payment vehicle is a voluntary donation and, therefore, the sampling frame can be the dwelling units within a city. As the survey method in this study, the valuation workshop method is employed, which decided the sampling frame should be a list of smaller populations. Because the survey will be made in Beijing, Guangzhou, and Nanchang, located in the Poyang Lake region, all the adult residents in the urban area of the three cities could be the sampling frame.

4.4.3 Sampling Design

Given the target population identified for the valuation study, an appropriate sample that represents this population must be obtained. The sample is the subset of the target population to whom the survey will be administered. A sample is used to save the money and time that would be otherwise required to survey the entire population (Bateman et al. 2002a). Therefore, the aim is that survey results of the sample can be extrapolated to the entire population. That is, the sample should be both representative and large enough to produce a sufficiently precise estimate of mean or median WTP value.

A probability sample has the characteristic that every element in the population has a known, non-zero probability of being included in the sample. A non-probability sample is a sample based on a sampling plan that does not have this feature. In probability sampling, estimates of population parameters can be constructed and, thereby, the validity and reliability of the resulting estimates can be evaluated, which cannot be finished with the non-probability sample (Levy and Lemeshow 1999). In Table 4.1, the categories of probabilistic design and non-probabilistic design are listed.

Probability sampling was abandoned in this study because it is often a time-consuming and expensive procedure (Levy and Lemeshow 1999). Furthermore, it is also not feasible in this study, because every element does not have a known chance of being selected in this survey under the condition that a limited sample size is selected from huge populations.

Meanwhile, non-probability samples are used quite frequently in economics and politics, especially in market research and public opinion surveys. *Judgment sampling* was abandoned in this study, as no insight can be gained mathematically concerning the reliability of the resulting estimates. In a CVM study, testing the reliability of estimate results mathematically is an essential step. *Quota sampling* was not applied in this study since very biased estimates are highly likely to be collected. Finally, *convenience sampling* was applied in this survey

Convenience sampling was chosen for various reasons: (1) It is the most effective method for research, in terms of time, effort, and money, which is an important factor in this study. (2) It is the least rigorous technique, involving the selection of the most accessible subjects. This survey aims at the most general population and does not have any requirement on participants. (3) It is the most prevalent sampling design and there is an element of convenience sampling in many qualitative studies (Marshall 1996). The main disadvantage of this sampling design is that the control of the selection process is very limited so that there is no valid basis for the inductive reference from the sample data to the population of interest (Anderson 2001). However, a formal inductive reference concerning the population of interest is not the focus of this study. A representative sample cannot be achieved when using a group-based approach (see Sect. 4.5), as the sample size is small. This implies that the estimate results cannot be used for the general population. The goal of this study is to make statements about the features of the sample itself. Therefore, convenience sampling was selected as the sampling design in our study.

Table 4.1 Taxonomy of sampling designs

		Definition	Features/Comments
Probabilistic designs	Simple random-sampling	Every element of the sample frame is given an equal chance of being selected	
	Systematic sampling	One selects every *k*th element from a randomly ordered population frame	
	Stratified sampling	The sample frame population is divided into distinct subpopulations, or strata. A separate and independent sample is selected with each stratum, using random sampling with either the same sampling fraction for each of the strata (proportionate stratification) or different sampling fractions (disproportionate stratification). The data are used to develop separate within–stratum estimates. Finally, the separate stratum estimates are combined (weighted) to form an overall estimate for the entire population	The strata are themselves of independent interest; one is free to over-sample surfers to obtain a sufficient sample size for separate analysis; variances differ between strata or survey cost are different for different strata, which can greatly increase efficiency for a given overall sample size or survey budget. Finally, the creation of strata permits the use of different sample designs for different portions of the population
	Cluster sampling	The population is divided into a set of groups or clusters but one selects only a random sample of the clusters. Strictly defined, cluster sampling involves sampling all the elements within the selected clusters, but the term is also used more loosely to cover multistage sampling, in which one selects only a random sample of the elements within the selected clusters	It is generally more convenient and more economical than simple random sampling; it is also attractive when no overall sample frame is available. It is common for CVM survey of large populations
Non-probabilistic designs	Convenience samples	One assembles a sample randomly, at the convenience of the researcher and with the minimal control over the selection	Most elementary and least satisfactory
	Judgment samples	A panel of respondents judged to be representative of the target population is assembled	Controlled but non-random. No insight can be gained mathematically concerning the reliability of the resulting estimates
	Quota samples	Selection is controlled in principle by the interviewers, who are directed to ensure that the samples they survey contain given proportions of various types of respondents, designed to reflect the population of interest	It emerged as the preferred sampling method in political opinion polls and market research in the 1930s. It is highly likely that such a method of selecting a sample can lead to estimates that are very biased

Source: Bateman et al. (2002a), Levy and Lemeshow (1999)

Table 4.2 Sample size distribution

Study site	Sample size	Expected valid sample size
Nanchang	80	50
Guangzhou	80	50
Beijing	80	50

4.4.4 *Sample Size*

Choosing the sample size is a balancing of cost and precision. The trade-offs involved in the selection of a sampling approach and choosing a survey mode culminate in the determination of a sample size. In this study, it was decided that the sample size will be 240, 80 samples in each city (see Table 4.2).

The sample size was determined on the basis of the following arguments:

Firstly, a valuation workshop, that is, a group-based approach, was applied in the survey, which will be clarified in Sect. 4.5. This survey mode has the consequence that the sample size cannot be very large, which is a common disadvantage of all group-based approaches. However, more information can be collected from each respondent, which allows the smaller sample size to be employed.

Secondly, especially in the case of non-use values, sample quality is more important than sample size. It is easier to guarantee a high sample quality with a relatively smaller sample size. More time and assistance can be spent per respondent when the sample size is small. As a result, respondents can be well-instructed and answers can be given immediately when they have any questions.

Thirdly, the survey budget may impose an absolute limitation on the size of the sample. Some reimbursement should be paid to participants. On the one hand, their efforts and time can be rewarded. On the other hand, money can be a motive for people to join in the survey. When the survey budget is limited, the sample size should be planned under the budget constraint.

A final argument is that this kind of survey is very new in China. The method was transferred successfully to the Chinese. This can also be a good foundation for future bigger surveys.

In conclusion, taking into account all the mentioned factors, such as survey mode, survey budget and time restriction, guarantee the precision of estimates etc., the sample size was determined to be 240 individuals. Among these factors, the survey mode is of utmost importance. In the subsequent final section of this chapter, the survey method is explained in more detail.

4.5 Survey Method

The main objective of this section is to review different survey methods for CV and discuss their advantages and disadvantages. Given the complexity of porpoise conservation, budget and time constraints, a proper survey method is selected.

The section begins with presenting and comparing traditional survey methods (Sect. 4.5.1). It then turns to group-based approaches (Sect. 4.5.2). Finally, the methods are compared and the method most suitable for the YFP survey will be chosen (Sect. 4.5.3).

4.5.1 Traditional Survey Methods

CV is traditionally applied through face to face interviews, mail, and telephone interview carried out by professionally trained staff. Among the three approaches, the face to face interview method is highly recommended by the influential NOAA report (Arrow et al. 1993).

4.5.1.1 Face to Face Interview

The face-to-face interview is one of the oldest forms of data collection in surveys, and it has evolved from a brief and simple inquiry in the 1930s to a complex and highly sophisticated research instrument (Rossi et al. 1983; Smith 1987). The face-to-face interview is the mode in which an interviewer administers a structured or partly structured questionnaire to a respondent within a limited period of time and in the presence of the respondent (De Leeuw 1992). The method is very widely used on account of its advantages.

Face-to-face surveys tend to obtain higher response rates compared with other survey methods, such as mail surveys or telephone surveys. Face-to-face interviews facilitate communication between respondents and interviewers. It is, therefore, easier for respondents to understand the questions and state true responses. What is more, it is embarrassing to refuse someone face-to-face, generally, which is very conventional in China. Consequently, response rates are higher than in mail or telephone surveys. Goyder (1987) collected data on 385 mail surveys, 112 face-to-face surveys, and 53 telephone surveys in the United States and Canada between 1930 and 1980. On average the response rate for the face-to-face interview was 67.3%, for the telephone interview 60.2%, and for the mailed questionnaire 58.4% (Goyder 1987).

A face-to-face interview is the most flexible form among all data collection methods. Structured or partly structured questionnaires and highly complex questionnaires can all be used because face-to-face explanation can compensate lacks in understanding. Respondents can be presented with all kinds of visual aid, ranging from simple response cards to posters or video clips. The interviewer can help respondents understand complex questions by providing in-time explanations; he can administer questionnaires with a large number of screen questions, control question sequence, and examine answers on open issues (De Leeuw 1992). All these measures can help people master the given information.

The above advantages can be fully exploited under certain conditions when respondents have sufficient time and are willing to cooperate with interviewers.

However, this is an ideal situation that cannot be achieved at all times. There are still some disadvantages associated with the method. Firstly, it is optimal to conduct face-to-face interviews in a relaxed environment, for example, at home or in a tea bar. Respondents can calm down and read the questionnaire carefully. But in many areas, especially in Beijing, it is impossible to knock at respondents' door. Therefore, in China, a face-to-face interview cannot take place at home, but more possibly in a shopping mall, KFC, McDonald's, or some other similar places. However, in these public places, people often walk in a hurry, and then they hesitate to spare time to complete a questionnaire, which would negatively influence the quality of the results.

Secondly, geographically dispersed face-to-face interviews require the longest time. The whole process includes the travel time, waiting time, chatting time, and the time to fill in the questionnaires. Chatting time is necessary for making a passer-by believe that the interviewer is not a cheater and make the interviewee become interested in the survey. After the warm opening, the questionnaire survey is possible to start. It should take 1–2 h to finish one questionnaire at least.

In summary, the ideal situation for face-to-face interviews is quite difficult to be satisfied. Moreover, the limitation of time and budget confine the coverage of face-to-face interviews. Therefore, face-to-face interviews are not the best choice in this survey.

4.5.1.2 Mail Survey

When a mail questionnaire is used, a respondent receives a structured questionnaire and an introductory letter by mail; he/she answers the questions in her/his own time without any assistance from the researcher except for any written instructions in the questionnaire or in the enclosed letter; and he/she finally sends the questionnaire back (De Leeuw 1992).

In general, mail surveys require an explicit sampling frame of names and addresses. A list of mailing addresses can be sought from a marketing institute, but this is rather expensive. Another drawback of mail surveys is the limited control the researcher has over the choice of the specific individual within the household who, in fact, completes the survey (De Leeuw 1992). It is possible that only those people that are most likely to be home at the time the interviews are taken. Thus, unemployed homemakers and retired people are likely to be disproportionately represented (Dillman 1978). Thus, a proportional representation cannot be ensured.

The third drawback of mail survey is that it has the lowest response rate compared to face-to-face surveys and telephone interviews (Dillman 1978). A certain number of questionnaires might fail to reach the respondents because their mail addresses have changed. At the same time, people who are not the addressee receive mails and simply decide to discard them, rather than forward it to the correct addressee.

The fourth drawback of mail surveys is that it is the least flexible data collection technique because all questions must be presented in a fixed order and all respondents receive the same instruction without added assistance in individual cases (De Leeuw 1992). However, available visual cues can help compensate this disadvantage to some extent. Information booklets or product samples can be sent by mail accompanying a questionnaire for their evaluation. But in the case of YFPs, these booklets are probably not adequate for respondents to finish the complex task of valuating the porpoise conservation project.

A further drawback of mail surveys is that it is usually locked into a definite time interval of mailing dates with rigidly scheduled follow-ups (De Leeuw 1992). In general, when the questionnaires are sent to people, people are informed about the deadline of returning the surveys. During the interval, some people would like to finish the questionnaires at the latest moment, while others fail to return the questionnaire and miss the deadline. Several letters reminding respondents of the deadline should also be sent, but these reminders mails may annoy people.

However, mail surveys have been more and more widely used due to their advantages. Organizational and personnel requirements for a mail survey are the least demanding (De Leeuw 1992). Only a few interviewers have to be employed to sort the mail addresses and send letters because the material package, including the questionnaire and assistant booklets, are designed and prepared by the administer. Thus the interviewers are not required to be skilled in the standard interview technique. Because of a low response rate, the low degree of flexibility, and the complexity of the porpoise conservation project, it was decided that a mail survey is not suitable for this study.

4.5.1.3 Telephone Interview

In a telephone interview, the interviewer administers the questions from a structured questionnaire and within limited time via telephone calls (De Leeuw 1992). Since the 1970s, telephone interviews have become increasingly popular in government agencies and survey research firms (Lyberg and Kasprzyk 1991), which is much related to the increasing telephone coverage all over the world.

However, telephone coverage is not 100%. Certain population groups, such as the unemployed, are more difficult to be reached by telephone, which can lead to biased estimates, especially when these groups are target groups (Snijkers 1992). Furthermore, in developing countries, the telephone coverage is still rather lower. For instance, in China, the fixed telephone coverage is 23.6% and the mobile phone coverage is 56.3%, according to the official statistics of 2009 (Ministry of Industry and Information Technology of the People's Republic of China 2009). In the end, a telephone survey cannot reach the general population in China.

Furthermore, telephone interviews have lower response rates than face-to-face interviews and slightly higher response rates than mail surveys (De Leeuw 1992). In most situations, people hang up when they recognize that the caller is a stranger.

Finally, telephone interviews are less flexible than face-to-face interviews. The absence of visual cues during the interview is a big drawback, which limits the number of response categories to be used. With a lack of visual documents, such as pictures and videos, people can only get information through listening and then give their response. It is a great challenge for interviewers and interviewees to communicate clearly. The response category can be divided into three kinds of responses simply—"YES," "NO," and "DON'T KNOW."

Nevertheless, telephone interview also have advantages. A telephone survey is very easy and relatively low cost (Kalfs and Saris 1991). The cost is marginally higher than mail surveys but much lower than face-to-face interviews. The telephone directory is conveniently accessible and can be used as a sampling frame. What needs paying is only the phone fee. With the rapid development of telecommunication techniques, phone fees are decreasing, which also promotes the application of this technique. Expression skills are most important and most highly required in the telephone surveys. Generally speaking, when interviewers have excellent expression skills, they can complete the survey according to the well-structured survey.

In summary, the lack of visual aids leads to biased estimates and lower response rates, both of which weaken the application of the telephone survey. Low coverage of telephones in China is an important additional reason to abandon the telephone survey in this research. In addition to the methods mentioned in this section, which are all based on interviewing individuals, there are methods where the information about the environmental good to be evaluated is given within a group-based meeting. In the following, I will explain these methods and describe their advantages and disadvantages.

4.5.2 Deliberative Group-Based Approaches

Deliberation refers either to a particular sort of discussion—one that involves the careful and serious weighting of reasons for and against some suggestion—or to an interior process by which an individual weighs up reasons for and against courses of action (Fearon 1998). Collective "problem-solving" discussion is viewed as the critical element of deliberation, to allow individuals with different backgrounds, interests and values to listen, understand, potentially persuade, and ultimately come to more reasoned, informed, and public-spirited decisions (Abelson et al. 2003). Deliberative features have been incorporated into a broad grouping of methods that include (1) citizens' juries/value juries, (2) market stalls, and (3) the valuation workshops. In the three group-based methods, participants are provided with information about the issue being considered. They are encouraged to discuss the provided information and then come to final statements.

4.5.2.1 Citizens' Jury

The Citizens' Jury was developed in the early 1970s in Germany and the USA. In Germany, the concept was shaped in the form of "citizen panels," invented by Dienel, while in the USA, it was Crosby who pioneered the process (Dienel and Renn 1995; Crosby 1995).

A Citizens' Jury consists of a small group of residents, who are selected to represent the general public rather than any particular interest group or sector. The small group of citizens meets on purpose to consult or confer formally upon a policy issue (Stewart et al. 1994). Citizen Juries usually consist of around 12–24 citizens, who are drawn from the relevant population as "symbolically representatives." Normally, the discussion process lasts over 2 or 3 days (Coote and Lenaghan 1997). The jury members listen to witnesses who present evidence on the issue, they ask questions to these witnesses, and they decide on an agreed preferred course of action (Kenyon et al. 2001). A Citizens' Jury allows participants' opinions and preferences to be expressed and considered. Therefore, every juror can have an equal impact on the final recommendations (Crosby 1995).

There are two potential roles of Citizens' Jury on environmental valuation of public goods. Firstly, it could be asked to recommend the maximum payment from citizens that would be reasonable to provide the public good at issue. This monetary amount could then be used along with the other value information to help managers reach a decision. Secondly, the jury could be asked to recommend an alternative among the management options under consideration, in light of the physical and economic information presented about the alternatives and in view of the jury's undetermined judgment of the social value of the public good (Brown et al. 1995).

A Citizens' Jury consists of a project director, who is responsible for overseeing the entire project, the project staff, an advisory committee and a working group. The function of the working group is to ensure that the project is planned in such a way as to accomplish the needs and objectives of the sponsors, while the project staff ensures that the integrity of the Citizens' Jury process is maintained (The Jefferson Center 2004).

The important advantages of a value jury are representative, objective, well-informed, formal, cost-effective in appropriate applications, and likely to adopt sound common sense. Firstly, members of a value jury would be chosen randomly. Secondly, the sampling frame of the jurors is an agency for the larger society, including both public and private interests, and not advocacy of a specific interest group. Thirdly, jury members become well-informed about the deliberated issue in the course of listening, asking questions, exchanging ideas during the information stage of the jury process. Fourthly, the value jury process is formal and systematic. Fifthly, a values jury is often more cost-effective than large-scale public surveys because less people are employed and less time is needed. Sixthly, the value jury probably bases its decisions on good common sense since every juror is equal to present recommendations (Brown et al. 1995).

The primary disadvantage of a values jury is that it consists of a small number of persons and, therefore, runs the risk of being unrepresentative of the rich pluralism of the public at large, which is similar for all deliberative methods. Furthermore, small groups of individuals who sit together for a considerable time may be susceptible to undue influence by unusually articulate or passionate members of the group. Possibly, it may fail to reach an agreement (Brown et al. 1995). What is more, to organize and implement one citizens' jury project is at great cost and time-consuming.

A Citizens' Jury is not fit for CV for several reasons: in contingent valuation, respondents give WTP as consumers in the hypothetical market of a public good.

However in the Citizens' Jury, jurors are chosen as agents of society who think of the welfare of the community when responding to environmental issues. Additionally, the aim of the Citizens' Jury is to obtain a recommendation that not even the jurors agree with; but, at least, nobody votes against it. In contrast, in contingent valuation, respondents are chosen at random and are required to give individual WTP bids. The mean value or the median value is the result that researchers really want to achieve. Furthermore, a Citizens' Jury allows participants' opinions and preferences to be expressed and registered regardless of their ability to pay, which is not permitted in CV. Therefore, a Citizens' Jury is more suitable for a policy issue rather than for a value judgment.

4.5.2.2 Market Stall

The market stall (MS) approach evolved out of recent experience with Citizens' Juries in environmental decision-making. The MS approach is designed to overcome the superficial nature of public surveys by providing participants with more time and information to decide on their WTP. It combines the desirable features of group techniques with the particular requirements of economic valuation and cost benefit analysis that are normally met from interview-based encounters. The approach was first applied to value the non-market benefits of wild goose conservation in Scotland. (MacMillan et al. 2002)

The MS approach involves between 10 and 15 participants attending two group meetings, between which there is approximately 1 week interval. The first meeting is primarily concerned with the presentation of relevant information about the proposed project described in an "Information Folder," and a detailed explanation of the hypothetical market in CV. Participants are given the opportunity to discuss any aspect of the project and to question the moderator of the group meeting. The group meeting concludes with a WTP question, which respondents answer confidentially in writing. During the week-long interval between the two group meetings, participants are asked to complete a daily diary in which they can record their thoughts and questions about the proposed project in the questionnaire. For example, participants can write down whether they would change their WTP and explain why. They can also write any suggestions and proposals that come into mind. If they get any new information from the Internet, TV, or newspapers, they can write it down. Further, if they have any questions related to the project, they can also record these and are ready to present them in the second group meeting. At the second group meeting, participants are given the opportunity to ask questions and to talk about any unresolved issue concerning the project. The WTP question is then readministered to participants. A debriefing exercise can be carried out to establish the extent to which participants understood the approach as a means of establishing the values they place on environmental goods or services (MacMillan et al. 2002).

The MS approach is designed in a way to generate higher levels of motivation, compared to conventional surveys. (1) Participants are paid for some money and, therefore, have a financial incentive. (2) In addition, participants attain more motivation when they feel their comments during the discussion are being taken seriously. They experience the feeling as the "owner" of the public goods and they feel that they can play an important role in environmental management, which has a lot to do with their daily life. Especially in China, people mainly focus on economic development and their environmental awareness is not highly prevalent. Generally speaking, they feel that environmental management and protection of endangered rare species are the government's business, though they suffer from environmental pollution and ecological degradation. With such a survey method, people's awareness can be aroused and participants could feel empowered as well as enjoy such an activity. Consequently, people can be highly motivated. (3) The informal and relaxed relationship between moderator and participants also seemed to develop trust for each other. Trust is very important for participants to express their true opinion based on their own WTP (Lienhoop and MacMillan 2007). In conclusion, participants' motivation can be stimulated most in the MS and then the answers given will be relatively more consistent and convincing.

Except for the high motivation, the MS addresses three essential inadequacies of conventional interviews: (1) it overcomes the "snapshot in time" weakness by providing participants with more time and information to decide their WTP; (2) it can deal with uncertainty because participants can take advantage of an informal setting where in-depth discussions with the moderator and other group members can take place; (3) the week-long interval between the two meetings provides the opportunity for participants to reevaluate their WTP following further thought, information searching, crucial household economic decisions, and discussions with family members and friends. Most importantly, more detailed deliberations can be facilitated to provide the decision-maker with more prosperous and complete pictures of peoples' perceptions of, and reactions to, the environmental issue (Kenyon and Hanley 2001; MacMillan et al. 2002).

However, sample sizes are typically small due to recruitment difficulty and financial constraints and, hence, would not be easy to expand the research to obtain a statistically valid and representative sample of the very general population (Lienhoop and MacMillan 2007). In addition, being a group-based approach, there is the risk that "group norms" might influence reported individual WTP. For example, certain participants may not wish to disagree with the rest of the group. Some participants may be too talkative and always try to impose their own opinions on other participants, which may also influence the WTP results (MacMillan et al. 2002). Polarization may also occur, whereby an individual may adopt a more extreme position only because he/she wants to distinguish himself or herself from the other participants (Isenberg 1986). There may be greater opportunity to think and act strategically in MS and then the strategic bias will accompany the estimation results (MacMillan et al. 2002). For instance, participants may use the 1-week interval and additional information to calculate the fair WTP rather than maximum WTP. Participants could also possibly overestimate the WTP in order to increase, or underestimate WTP so as to decrease the probability of environmental goods being supplied.

4.5.2.3 The Valuation Workshop

The valuation workshop is a hybrid of the CV and Citizens' Jury. It aims to build on the strengths of the CV and Citizens' Jury exercises, and contains elements of both. Particularly, it intends to combine the quantitative outputs of the CV with the participatory, deliberative, and preference construction aspects of the Citizens' Jury (Kenyon and Hanley 2001).

The valuation workshop is divided into three parts. The first part is to make a short introduction to enable the moderator and participants get to know each other. Then, the participants are given 20–30 min to individually complete the questions on environmental issues and socio-economic characteristics. If participants have any questions about the questionnaire during the period, they can resort to the moderator. The second part is to read through the information folder. Participants are free to ask any questions about the YFPs that they do not understand or they would like to know. They are also encouraged to join in the group discussion. In this part, respondents are supposed to gain overall knowledge of the YFPs, including biological features, distribution, populations, threats, and importance. It takes about 30–40 min. In the last and most important part of the valuation workshop, participants are asked to read the information about the hypothetical market and complete some further questions individually on it. Participants can talk about the advantages and disadvantages of the proposed project and offer a ranked list of pros and cons according to the importance of points in their opinion. Their concerns about the project should be made clear. They can also provide some suggestions and recommendations for the project. At the same time, if they experience difficulties in answering any questions, they are free to resort to the moderator. Definitely, participants should indicate their individual WTP and explain why they are willing to pay a certain amount of money. It takes about 1 h. Before participants return to the questionnaire, they are asked to check their WTP amounts again and to make sure that they do not want to change their answer. In sum, it is estimated that the workshop will last 2–2.5 h.

The first advantage of the valuation workshop is that people can become well-informed. During the workshop, respondents can assimilate information fully and consider the project from different perspectives, given the interaction with other participants. Furthermore, it provides additional information on the areas of concern, and does not focus merely on positive aspects of the project. Participants are asked not only to discuss issues which they thought were problematic, but also to offer solutions to these problems (Kenyon and Hanley 2001). Secondly, individuals have enough time to form and mature their answers and then give their final decision. Compared to other survey methods, such as face-to-face interview or telephone interview, individuals have more time to deal with questions. Maturation is derived from a process in which the combination of deliberation and enough time to think leads to doubts and thoughts to emerge, which can be incorporated into the analysis (Whittington et al. 1992). Finally, the valuation workshop is not at very high cost and does not take too much time. The cost includes the reimbursement for participants, rent for meeting rooms and travel cost. A workshop with 10–15 participants lasts 2–3 h, which is very efficient, unlike face-to-face interviews, where more than half an hour is needed for one interviewee at least.

However, to organize workgroups is not easy because it lasts 2–2.5 h, which is very long for participants. Some participants might feel bored by the long discussion and they might not take the discussion seriously. Additionally, it is possible that some participants can be influenced by the passionate participants, which is a common shortcoming of all deliberative group-based approaches.

4.5.3 Comparison and Choice of Survey Methods

After having described and discussed the different survey methods, they should be compared in the final section, and the choice of the survey method for the YFPs will be explained.

First, a group-based approach is associated with relatively low cost, compared with other survey methods. The cost includes time costs, travel costs, and recruitment costs.

1. An interview, similar to the telephone interview, takes between 30 min and 1 h. Mail surveys and web-based surveys have a deadline for finishing the questionnaire. Here, the time span is different, and it may range from several weeks to several months. But the time how long respondents think about the problem is never known by interviewers. It depends on respondents' available time and their preferences. Some respondents may spend lots of time on thinking about the issue, while other respondents may give the answer at the latest moment and rarely think about the question seriously. In group-based approaches, one group takes 2–3 h and in this time span 10–15 questionnaires can be completed. By contrast, 5–10 h are needed for finishing 10–15 copies of questionnaire by some of the other survey methods.
2. Concerning travel cost, telephone interviews and mail surveys do not require travelling. Interviewers need to pay, travel costs in face-to-face interviews. Both interviewers and interviewees should be paid for travelling, also, in the group-based interviews because they need to come together at a certain place.
3. Regarding the recruitment costs, telephone interviews and mail interviews do incur costs for interviewees. In face-to-face interviews, it is possible to pay for interviewees. Normally, the price is different in different survey areas, which is much associated with the local income level. Usually, the higher the income level that respondents can achieve, the higher the required payment is. In group-based surveys, all the participants should also be paid. Here, the price is even little higher than in personal interviews. However, when the sample size is limited, the total recruitment cost is not very high.

Secondly, group-based surveys should provide the highest benefits. The benefits are measured by the validation of the estimation results. According to existent standards, the validation of estimation results is significantly positive related to respondents' motivation and information on the issue (Ajzen et al. 1996).

Obviously, the participants in the group-based surveys have the greatest motivation for two reasons. On the one hand, respondents are higher paid in group-based surveys compared to other survey methods—the monetary incentive is highest. On the other hand, participants can gain the feeling of being empowered in group-based surveys. Consequently, they could have higher motivation in the group-based interviews than in other forms of interviews.

In addition to higher motivation in groups, participants can be better informed in group-based surveys. (1) They have enough time to read the information folder and can ask whatever they want to know from the moderators. (2) Participants can share information with one another during the group discussion. (3) The relaxing environment can make individuals concentrate on questions. Individuals are instructed by the moderator to consider the project thoroughly. (4) Respondents have enough time to format and mature their answers. All these aspects can improve the validity of the survey results.

In addition, individuals in China are unfamiliar with CVM and the YFPs, which is also an important argument for selecting a group-based approach. CVM is mainly applied in developed countries and is rather new in China. Until now there have been very few CVM studies. For example, Kontoleon and Swanson (2003) conducted a CVM study to elicit WTP for the conservation of the giant pandas and their habitat. Their study is based on informal interviews only to foreign tourists in Beijing, China. Xu et al. (2003) applied CV in China to measure the TEV of restoring ecosystem services with in-person interviews in the Ejina region, China. Jim and Chen (2006) conducted CV of urban green spaces using in-person interviews in Guangzhou, China. Hammitt and Zhou (2006) valuated the economic value of air-pollution-related health risks by in-person interviews in China. Wang and Mullahy (2006) conducted a CVM study to estimate WTP for reducing fatal risk by improving air quality with in-person interviews in Chongqing, China. Yang et al. (2008) applied CVM to assess the ecosystem service value of constructed wetlands using face-to-face interviews in Hangzhou, China. Jin et al. (2008) valued the black-faced spoonbill conservation in Macao using a CVM study through a household drop-off survey mode.[6] Wang and Zhang (2009) estimated WTP for air quality by CV in Ji'nan, China through a series of face-to-face interviews. All existent studies only involved a limited number of people. Nevertheless, one can say that CVM is rarely applied and the Chinese are rather unfamiliar with the method. Moreover, many Chinese are not familiar with the YFPs.

[6] In the *household drop-off* survey, a researcher goes to the respondent's home or business and hands the respondent the questionnaire. In some cases, the respondent is asked to mail it back or the interview returns to pick it up. This approach attempts to blend the advantages of the mail survey and the group administered questionnaire. Like the mail survey, the respondent can work on the questionnaire in private, when it is convenient. Like the group administered questionnaire, the interviewer makes personal contact with the respondent—they do not just send an impersonal survey instrument. And, the respondent can ask questions about the study and gets clarification on what is to be done. Generally, this would be expected to increase the percentage of people who are willing to respond.

After having decided for a group-based approach, one should be chosen out of the three group-based approaches as the best-suited for the YFP research. According to the above analysis, the Citizens' Jury is more suited to a policy issue, which is not consistent with the research goal in this study. In this case study, the aim is to value the YFPs by asking respondents their WTP. Therefore, the Citizens' Jury is rejected in this study. Thus, preference is given to MS and to valuation workshop. Comparing these two methods, respondents have more time to think and can become more informed on the deliberated issue in the MS. Respondents can search additional information from all kinds of media during the week interval. Also, they could reconsider their WTP answers carefully. In this study, many people are not familiar with or have never even heard of the YFPs. If individuals could spend more time on collecting information about porpoises and take the chance to talk about the questionnaire with their family or friends, it is very helpful for them to construct a preference. From this perspective, it can be judged that MS is better than valuation workshop.

However, the feasibility of survey methods is also an important factor and should also be taken into account. The survey will be carried out in China, a fast-developing country. In China, people live at a fast pace, especially in megacities. Everybody seems to be very busy. Only retired old people have more time. If MS is applied, the possibility seems to be very small that individuals would like to do additional research on the topic in the week interval. In addition, the MS requires participants to write a daily dairy for 1 week, which is obviously too demanding for respondents in China. Furthermore, two group meetings will increase the difficulty of recruitment. Moreover, MS of two meetings should require higher recruitment costs to be paid for participants than valuation workshops that only have one meeting. Finally, time restraint is an important reason for selecting the valuation workshop as an adequate method. MS requires much more time than valuation workshops. All these arguments led us to the conclusion, valuation workshops are recommended in this survey in China.

References

Abelson, J., Forest, P. G., Eyles, J., Smith, P., Martin, E., & Gauvin, F. P. (2003). Deliberations about deliberative methods: issues in the design and evaluation of public participation processes. *Social Science & Medicine, 57*, 239–251.

Ajzen, I., Brown, T. C., & Rosenthal, L. H. (1996). Information bias in contingent valuation: Effects of personal relevance, quality of information, and motivational orientation. *Journal of Environmental Economics and Management, 30*, 43–57.

Anderson, D. R. (2001). The need to get basics right in the wildlife field studies. *Wildlife Society Bulletin, 29*, 1294–1297.

Arrow, K., Solow, R., Portney, P. R., Learner, E. E., Radner, R., & Schuman, H. (1993). Report of the NOAA panel on contingent valuation. *Federal Register, 58*, 4601–4614.

Bateman, R. P., Carey, M., Moore, D., & Prior, C. (1993). The enhanced infectivity of Metarhizium flavoviride in oil formulations to desert locusts at low humidities. *Annals of Applied Biology, 122*, 145–152.

Bateman, I. J., Brainard, J. S., & Lovett, A. A. (1995). *Modeling woodland recreation demand using geographical information systems: A benefit transfer study* (CSERGE WP GEC 95–06). Norwich: School of Environmental Sciences, University of East Anglia. Retrieved from http://www.cserge.ac.uk/sites/default/files/gec_1995_06.pdf

Bateman, I. J., Langford, I. H., Jones, A. P., & Kerr, G. N. (2001). Bound and path effects in double and triple bounded dichotomous choice contingent valuation. *Resource and Energy Economics, 23*, 191–213.

Bateman, I. J., Carson, R. T., Day, B., Hanemann, M., Hanley, N., Hett, T., Jones-Lee, M., Loomes, G., Mourato, S., Özdemiroglu, E., Pearce, D., Sugden, R., & Swanson, J. (2002a). *Economic valuation with stated preference techniques: A manual.* Cheltenham: Edward Elgar.

Boxall, P. C., Adamowicz, W., Swait, J., Williams, M., & Louviere, J. (1996). A comparison of stated preference methods for environmental valuation. *Ecological Economics, 18*, 243–253.

Brown, T. C., Peterson, G. L., & Tonn, B. E. (1995). The values Jury to aid natural resource decisions. *Land Economics, 71*, 250–260.

Bunge, M. (1974). Treatise on basic philosophy, Vol. I Semantics I: Sense and Reference. Dordrecth-Boston: Reidel Publishing Co.

Champ, P. A., Bishop, R. C., Brown, T. C., & McCollum, D. W. (1997). Using donation mechanisms to value nonuse benefits from public goods. *Journal of Environmental Economics and Management, 33*, 151–162.

Coote, A., & Lenaghan, J. (1997). *Citizen's juries: Theory into practice.* London: Institute for Public Policy Research (IPPR).

Crosby, N. (1995). Citizen juries: One solution for difficult environmental questions. In O. Renn, T. Webler, & P. Wiedemann (Eds.), *Fairness and competence in citizen participation* (pp. 157–174). Boston: Kluwer Academic Publishers.

Cummings, R. G., Brookshire, D., & Schulz, W. D. (1986). *Valuing environmental goods: An assessment of the contingent valuation methodology*. Savage: Rowman & Littlefield Publishers.

Cummings, R. G., Elliott, S., Harrison, G. W., & Murphy, J. (1997). Are hypothetical referenda incentive compatible? *Journal of Political Economy, 105*, 609–621.

De Leeuw, E. D. (1992). *Data quality in mail, telephone, and face-to-face surveys.* Amsterdam: TT Publications. Retrieved from http://www.eric.ed.gov/PDFS/ED374136.pdf

Desvousges, W. H., Smith, V. K., & Fisher, A. (1987). Option price estimates for water quality improvements: A contingent valuation study for the monongahela river. *Journal of Environmental Economics and Management, 14*, 248–267.

Desvousges, W. H., Johnson, F. R., Dunford, R. W., Boyle, K. J., Hudson, S. P., & Wilson, N. (1993). Measuring natural resource damages with contingent valuation: Tests of validity and reliability. In J. A. Hausman (Ed.), *Contingent valuation: A critical assessment* (pp. 91–159). Amsterdam: North Holland.

Dienel, P., & Renn, O. (1995). Planning cells: A gate to fractal mediation. In O. Renn, T. Webler, & P. Wiedemann (Eds.), *Fairness and competence in citizen participation* (pp. 117–140). Boston: Kluwer Academic.

Dillman, D. A. (1978). *Mail and telephone surveys: The total design method.* London: Wiley.

Duffield, J. W., & Patterson, D. A. (1992, February 23–27). Field testing existence values: Comparison of hypothetical and actual cash transaction values. In R. B. Rettig (Comp.), *Benefits and costs in natural resource planning, fifth interim report.* South Lake Tahoe/Corvallis: Oregon State University.

Everitt, B. S. (2002). *The Cambridge dictionary of statistics.* Cambridge: Cambridge University Press.

Fearon, J. D. (1998). Deliberation as discussion. In J. Elster (Ed.), *Deliberative democracy* (pp. 44–68). Cambridge: Cambridge University Press.

Garrod, G., & Willis, K. G. (1999). *Economic valuation of the environment: Methods and case studies.* Cheltenham: Edward Elgar Publishing.

Goyder, J. (1987). *The silent minority: Non-respondents on sample surveys.* Cambridge: Perseus Publishing.

Green, C., & Tunstall, S. (1999). A psychological perspective. In I. J. Bateman & K. G. Willis (Eds.), *Valuing environmental preferences* (pp. 207–257). Oxford: Oxford University Press.

Hammitt, J., & Zhou, Y. (2006). The economic value of air-pollution-related health risks in China: A contingent valuation study. *Environmental and Resource Economics, 33*, 399–423.

Harris, C. C., Driver, B. L., & McLaughlin, M. J. (1989). Improving the contingent valuation method: A psychological approach. *Journal of Environmental Economics and Management, 17*, 213–229.

Isenberg, D. J. (1986). Group polarization: A critical review and meta-analysis. *Journal of Personality and Social Psychology, 50*, 1141–1151.

Jim, C. Y., & Chen, W. Y. (2006). Recreation – Amenity use and contingent valuation of urban greenspaces in Guangzhou, China. *Landscape and Urban Planning, 75*, 81–96.

Jin, J. J., Wang, Z. S., & Liu, X. M. (2008). Valuing black-faced spoonbill conservation in Macao: A policy and contingent valuation study. *Ecological Economics, 68*, 328–335.

Kalfs, N., & Saris, W. E. (1991). Mode effects in time diary research. *Kwantitatieve Methoden, 37*, 65–86.

Kenyon, W., & Hanley, N. (2001). *Economic and participatory approaches to environmental evaluation* (Discussion Paper 00–15). Glasgow: Economics Department, University of Glasgow. Retrieved from https://dspace.gla.ac.uk/bitstream/1905/534/1/2000_15.pdf

Kenyon, W., Hanley, N., & Nevin, C. (2001). Citizens' juries: An aid to environmental valuation? *Environment and Planning C: Government and Policy, 19*, 557–566.

Kerr, G. N. (1996). Probability distributions for dichotomous choice contingent valuation (CSERGE GEC Working Paper). Norwich: University of East Anglia. Retrieved from http://www.prototype2010.cserge.webapp3.uea.ac.uk/publications/cserge-working-paper/gec-1996-08-probability-distributions-dichotomous-choice-contingent-valuation

Kitzinger, J. (1994). The methodology of focus groups: The importance of interactions between research participants. *Sociology of Health & Illness, 16*, 103–121.

Kitzinger, J. (1995). Qualitative research: Introducing focus groups. *British Medical Journal, 311*, 299–302.

Kontoleon, A., & Swanson, T. (2003). The willingness to pay for property rights for the giant panda: Can a charismatic species be an instrument for nature conservation? *Land Economics, 79*, 483–499.

Krueger, R. A., & Casey, M. A. (2000). *Focus groups: A practical guide for applied research* (3rd ed.). Thousand Oaks: Sage Publications.

Levy, P. S., & Lemeshow, S. (1999). *Sampling of populations: Methods and applications* (3rd ed.). New York: Wiley-Interscience.

Lienhoop, N., & MacMillan, D. (2007). Contingent valuation: Comparing participant performance in group-based approaches and personal interviews. *Environmental Values, 16*, 209–232.

Loomis, J. B., & Gonzalez-Caban, A. (1997). Comparing the economic value of reducing fire risk to spotted owl habitat in California and Oregon. *Forest Science, 43*, 473–482.

Loomis, J., & Larson, D. (1994). Total economic values of increasing gray whale populations: Results from a contingent valuation survey of visitors and households. *Marine Resource Economics, 9*, 275–286.

Lyberg, L., & Kasprzyk, D. (1991). Data collection methods and measurement errors: An overview. In P. Biemer, R. M. Groves, L. E. Lyberg, N. A. Mathiowetz, & S. Sudman (Eds.), *Measurement errors in surveys* (pp. 237–258). New York: Wiley.

MacMillan, D., Philip, L., Hanley, N., & Alvarez-Farizo, B. (2002). Valuing the non-market benefits of wild goose conservation: A comparison of interview and group-based approaches. *Ecological Economics, 43*, 49–59.

Mathieu, L. F., Langford, I. H., & Kenyon, W. (2003). Valuing marine parks in a developing country: A case study of the Seychelles. *Environment and Development Economics, 8*, 373–390.

Marshall, M. N. (1996). Sampling for qualitative research. *Family Practice, 13*, 522–525.

Ministry of Industry and Information Technology of the People's Republic of China. (2009). *Statistics of China telecommunication in 2009*. Retrieved from http://www.miit.gov.cn/n11293472/n11293832/n11294132/n12858447/13011909.html

Mitchell, R. C., & Carson, R. (1981). *An experiment in determining willingness to pay for national water quality improvements*. Washington, DC: Draft Report to the US Environmental Protection Agency. Retrieved from http://yosemite.epa.gov/ee/epa/eerm.nsf/vwAN/EE-0011-01.pdf/$file/EE-0011-01.pdf

Mitchell, R. C., & Carson, R. T. (1984). *A contingent valuation estimate of national freshwater benefits*. Washington, DC: Technical Report to the U.S. Environmental Protection Agency. Retrieved from http://yosemite.epa.gov/ee/epa/eerm.nsf/vwAN/EE-0219-01.pdf/$file/EE-0219-01.pdf

Mitchell, R. C., & Carson, R. T. (1989). *Using surveys to value public goods: The contingent valuation method*. Baltimore: Johns Hopkins Press.

Morrison, M. D., Blamey, R. K., & Bennett, J. W. (2000). Minimising payment vehicle bias in contingent valuation studies. *Environmental and Resource Economics, 16*, 407–422.

Navrud, S. (1992). *'Willingness to Pay for Preservation of Species – an Experiment with Actual Payments', in pricing the European environment*. New York: Oxford University Press.

Nocera, S., Bonato, D., & Telser, H. (2002). The contingency of contingent valuation: How much are people willing to pay against Alzheimer's disease? *International Journal of Health Care Finance and Economics, 2*, 219–240.

Randall, A., Hoehn, J. P., & Tolley, G. S. (1981). *The structure of contingent markets: Some empirical results*. Paper presented at the annual meeting of the American Economic Association, Washington, DC.

Reaves, D. W., Kramer, R. A., & Holmes, T. P. (1999). Does question format matter? Valuing and endangered species. *Environmental and Resource Economics, 14*, 365–383.

Rossi, P. H., Wright, J. D., & Anderson, A. B. (1983). Sample surveys: History, current practice, and future prospects. In P. H. Rossi, J. D. Wright, & A. B. Anderson (Eds.), *Handbook of survey research* (pp. 1–20). San Diego: Academic.

Rowe, R. D., Schulze, W. D., & Breffle, W. S. (1996). A test for payment card biases. *Journal of Environmental Economics and Management, 31*, 178–185.

Schuman, H. (1996). The sensitivity of CV outcomes to CV survey methods. In D. J. Bjornstad & J. Kahn (Eds.), *The contingent valuation of environmental resources: Methodological issues and research needs*. Cheltenham: Edward Elgar.

Seip, K., & Strand, J. (1992). Willingness to pay for environmental goods in Norway: A contingent valuation study with real payment. *Environmental and Resource Economics, 2*, 91–106.

Smith, T. W. (1987). The art of asking questions-1936–1985. *Public Opinion Quarterly, 51*, 95–108.

Snijkers, G. J. M. E. (1992). Computer assisted interviewing: Telephone or personal? In A. Westlake (Ed.), *Survey and statistical computing* (pp. 137–146). Amsterdam: Elsevier Science Publishers.

Spencer, M. A., Swallow, S. K., & Miller, C. J. (1998). Valuing water quality monitoring: A contingent valuation experiment involving hypothetical and real payments. *Agricultural and Resource Economics Review, 27*, 23–42.

Steel, R. G. D., & Torrie, J. H. (1960). *Principles and procedures of statistics*. New York/Toronto/London: McGraw-Hill.

Stewart, J., Kendall, E., & Coote, A. (1994). *Citizens' juries*. London: Institute for Public Policy Research.

Swallow, B. M., & Woudyalew, M. (1994). Evaluating willingness to contribute to a local public good: Application of contingent valuation to tsetse control in Ethiopia. *Ecological Economics, 11*, 153–161.

The American Psychological Association (APA), American Educational Research Association (AERA), and the National Council on Measurement in Education (NCME). (1974). *Standards for educational and psychological tests and manuals*. Washington, DC: American Psychological Association.

The Jefferson Center. (2004). *Citizens' Jury handbook*. Retrieved from http://www.epfound.ge/files/citizens_jury_handbook.pdf

Tisdell, C., & Wilson, C. (2006). Information, wildlife valuation, conservation: Experiments and policy, contemporary economic policy. *Western Economic Association International, 24*, 144–159.

Wang, H., & Mullahy, J. (2006). Willingness to pay for reducing fatal risk by improving air quality: A contingent valuation study in Chongqing, China. *Science of the Total Environment, 367*, 50–57.

Wang, Y., & Zhang, Y. S. (2009). Air quality assessment by contingent valuation in Ji'nan, China. *Journal of Environmental Management, 90*, 1022–1029.

Whitehead, J. C. (2002). Incentive incompatibility and starting-point bias in iterative valuation questions. *Land Economics, 78*, 285–297.

Whitehead, J. C., Blomquist, G. C., Hoban, T. J., & Clifford, W. B. (1995). Assessing the validity and reliability of contingent values: A comparison of on-site users, off-site users, and non-users. *Journal of Environmental Economics and Management, 29*, 238–251.

Whittington, D., Smith, V. K., Okorafor, A., Okore, A., Liu, J. L., & McPhail, A. (1992). Giving respondents time to think in contingent valuation studies: A developing country example. *Journal of Environmental Economics and Management, 22*, 205–225.

Xu, Z. M., Cheng, G., Zhang, Z., Su, Z., & Loomis, J. (2003). Applying contingent valuation in China to measure the total economic value of restoring ecosystem services in Ejina region. *Ecological Economics, 44*, 345–358.

Yang, W., Chang, J., Xu, B., Peng, C., & Ge, Y. (2008). Ecosystem service value assessment for constructed wetlands: A case study in Hangzhou, China. *Ecological Economics, 68*, 116–125.

Chapter 5
Results

Abstract In the chapter, collected information and data are presented and analyzed, such as sample characteristics and respondents' attitudes on the protection of YFPs. Respondents' socioeconomic characteristics are recorded, which serve further analysis of WTP bids—whether the WTP bids offered by the respondents are valid. In addition, based on the statements explaining how to decide the WTP amount, WTP bids are judged to be biased or valid. Furthermore, it also investigates the distribution of valid WTP bids, provides the descriptive statistics for data in the three cities, tests the validity of WTP responses statistically, compares the mean WTP in the three cities to test the distance effect, and analyzes mean WTP in China and Germany.

The result is that mean WTP in Beijing > mean WTP in Nanchang > mean WTP in Guangzhou, while they do not differ from each other statistically. Additionally, it is found out that dependent WTP bids of the respondents in Beijing have a significant relationship with the determinants *Edu.Degree, INCOME*, *BEQUEST,* and *Econ. AndEnvi.* at the 0.00 level ($R^2 = 0.434$). Dependent WTP estimates in Guangzhou are significantly related to the variables *Envi.Group, SIGHTSEEING, BIODIVERSITY,* and *INCOME* at the 0.00 level ($R^2 = 0.434$). Around 44% variance of WTP bids in Nanchang can be well-explained by the variables *Income*, *Fam.Size*, and *Envi.Group* at the 0.00 level. Last, the mean WTP in Germany is significantly larger than that in China.

In this chapter, the data obtained from the workshops are summarized and analyzed in the following way:

- Socioeconomic characteristics of the samples (Sect. 5.1 and 5.4.1).
- Evaluation of respondents' attitude (Sect. 5.2).
- Validity assessment of the WTP bids (Sect. 5.3).
- Descriptive statistical results for WTP estimates (Sect. 5.4.2).
- Theoretical validity tests of WTP (Sect. 5.4.3).
- Comparison of WTP bids in Beijing, Guangzhou, and Nanchang to test the distance effects (Sect. 5.4.2.1).
- Comparison of WTP means in China and Germany (Sect. 5.4.2.2).

Y. Dong, *Contingent Valuation of Yangtze Finless Porpoises in Poyang Lake, China*,
DOI 10.1007/978-94-007-2765-6_5, © Springer Science+Business Media Dordrecht 2013

Keywords WTP bids • Valid responses • Invalid responses • ANOVA • Multiple regression analysis

5.1 Sample Characteristics

One scientific aim of this study was to test the relevance of the distance effects in WTP studies. Therefore, 21 workshops were carried out between October and November 2009 in Beijing, Guangzhou, and Nanchang. A total of 240 respondents showed up in the workshops.

In order to test the validity of WTP, it is necessary to check the representativeness of the population. The three subsamples were compared because the socioeconomic equality among them is needed for testing the distance effect. Therefore, respondents were asked to answer questions on gender, age, educational level, income, family size, environmental membership, job, and ethnicity. The characteristics of the population are summarized in this section. In addition, it is also necessary to test the sampling error by *chi*-square comparisons of distributions for socioeconomic variables between the sample and the population.[1]

5.1.1 Gender Distribution

The distribution of socioeconomic respondent characteristics in the overall sample shows that the representation of male and female participants is approximately equal in three cities (Beijing: male 50%, female 50%; Guangzhou: male 51.3%, female 48.8%; Nanchang: male 50%, female 50%) (see Table 5.1). In China, the men account for 51.5% of the whole population while the women account for 48.5% of the whole population (China POPIN 2008).

A comparison of socioeconomic characteristics of the sample with official population statistics in the three cities[2] was made by a *chi*-square test on gender distribution (see Table 5.2). All the p-values in the three cities are much higher than 0.05, H_0 is accepted and the results are insignificant. Therefore, it can be concluded that the three subsamples are representative in terms of gender (Box 5.1).

---

[1] The population is the entire set of individuals to which findings of the survey are to be extrapolated (Levy and Lemeshow 1999). In this study, it refers to the people in Beijing, Guangzhou and Nanchang.

[2] Gender proportions (male *VS.* female) in Beijing, Guangzhou and Nanchang are: 52.1% *VS.* 47.9 %, 51% *VS.* 49%, 52% *VS.* 48% respectively. (Data of the fifth population census, National Bureau of Statistics People's Republic of China 2001).

Table 5.1 Gender distribution

	Beijing ($n=80$)		Guangzhou ($n=80$)		Nanchang ($n=80$)	
	N	(%)	*N*	(%)	*N*	(%)
Male	40	50	40	50	41	51.3
Female	40	50	40	50	39	48.8

Box 5.1 *Chi*-Square Test

Chi-square is a statistical test commonly used to compare observed data with data we would expect to obtain according to a specific hypothesis, named null hypothesis.

$$P = \sum_{i=1}^{n} \frac{(O_i - H_{0i})^2}{H_{0i}},$$

where
P = the test statistic that asymptotically approaches a χ^2 distribution.
Oi = an observed frequency;
$H_0 i$ = an expected (theoretical) frequency, asserted by the null hypothesis;
n = sample size.
To sum up:

If the *p*-value is greater than 0.05, H_0 is accepted and the result is not significant;
If the *p*-value is less than 0.05 but greater than 0.01, H_0 is rejected and the result is significant beyond the 5% level.
If the *p*-value is less than 0.01, H_0 is rejected and the result is significant beyond 1% level.

As suggested by Edwards and Anderson (1987, p. 170), to compare sample demographics, a *chi*-square statistic can be used to test the equality of the distributions with respect to gender, education, age, and income. There are some other examples in CV research. Loomis and King (1994) tested the independence of overall response rate to a mail and telephone-mail survey, and Carson et al. (1995) tested whether the distribution of respondents' choices "for" and "against" an environmental program changed between two independent surveys conducted 2 years apart.

5.1.2 Age Distribution

The age groups are not equally represented in our sample. The young age groups, especially the age groups from 16 to 25 and from 26 to 35 are overrepresented, while the old age groups (above 55 years) are underrepresented (see Table 5.3).

Table 5.2 *Chi*-square test statistics of gender distribution

	Beijing	Guangzhou	Nanchang
Chi-square	0.141[a]	0.032[a]	0.018[a]
df[b]	1	1	1
Asymp. sig.[c]	0.707	0.858	0.893

[a]0 cells (0.0%) have expected frequencies <5. The minimum expected cell frequency is 40.0
[b]*df* is the abbreviation of the degrees of freedom. In statistics, the number of degrees of freedom is the number of values in the final calculation of a statistic that are free to vary (http://www.animatedsoftware.com/statglos/sgdegree.htm)
[c]*Asymp. sig.* Asymptotic Significance, it calculates the significance level of a test via likelihood ratio methods (http://pbil.univ-lyon1.fr/library/asypow/html/asypow.sig.html)

Table 5.3 Age distribution

		Beijing ($n=80$)		Guangzhou ($n=80$)		Nanchang ($n=80$)	
Age category		*N*	(%)	*N*	(%)	*N*	(%)
16–65	16–25	21	26.3	29	36.3	19	23.8
	26–35	33	41.3	44	55.0	23	28.8
	36–45	13	16.3	1	1.3	15	18.8
	46–55	8	10.0	3	3.8	18	22.5
	56–65	4	5.0	1	1.3	4	5
	>65	1	1.3	2	2.5	1	1.3

According to the population statistics in China in 2008, the people of the age group (from 0 to 14 years) account for 19% of the total population, the people of the age group (from 15 to 64 years) account for 72.7%, and the age group (above 65 years) accounts for 8.3% (National Bureau of Statistics People's Republic of China 2009).

A *chi*-square test was conducted for comparing the subsamples with the population statistics in the three cities.[3] Specifically, respondents of old age (above 65 years old) are underrepresented in the subsamples of Beijing and Nanchang. It would be better if more people of this age had attended the workshops in Beijing and Nanchang. It can be understandable because the subsample sizes, 60 per city, are too small, compared with the huge populations in the two cities (Beijing, 16.95 million; Nanchang: 4.91 million) (China Population and Employment Statistics Yearbook 2008) (Table 5.4).

[3] Data of age distribution from the fifth population census in China, 2001

Age	0–14 (%)	15–65 (%)	>65 (%)
Beijing	13.6	78.0	8.4
GuangDong Province (Capital: Guangzhou)	24.17	69.78	6.05
Jiangxi Province (Capital: Nanchang)	25.99	67.91	6.11

Source: The State Council of People's Republic of China (2001)

Table 5.4 *Chi*-square test of age distribution

	Beijing	Guangzhou	Nanchang
Chi-Square	6.616[a]	3.288[a]	4.952[a]
df	1	1	1
Asymp. sig.	0.010	0.070	0.026

Note: It was decided to reclassify age brackets into 16–65 and >65 to be in accordance with existent statistical data

[a]0 cells (0.0%) have expected frequencies less than 5. The minimum expected cell frequency is 5.6

Table 5.5 Individual monthly income distribution

Category	Income level (RMB)	Beijing ($n=80$)		Guangzhou ($n=80$)		Nanchang ($n=80$)	
		N	(%)	*N*	(%)	*N*	(%)
1	Less than 500	17	21.3	22	27.8	16	20.0
2	500–1,499	3	3.8	4	5.0	19	23.75
3	1,500–2,499	6	7.5	15	18.8	29	36.25
4	2,500–3,499	19	23.8	12	15.0	12	15.0
5	3,500–4,499	20	25	8	10.0	4	5.0
6	4,500–5,499	2	2.5	7	8.75	0	0
7	5,500–6,999	6	7.5	8	10.0	0	0
8	7,000–8,999	3	3.8	3	3.75	0	0
9	9,000–9,999	3	3.8	1	1.3	0	0
10	Over 10,000	1	1.3	0	0	0	0

Note: 1 RMB=0.109385 Euro (April, 2010)

5.1.3 Income Distribution

Individual monthly income is distributed evenly, though income groups (<500 RMB) are slightly overrepresented and the income groups (>4,500 RMB) are slightly underrepresented, especially in Nanchang (see Table 5.5). Specifically, mean income level of respondents is 4.15 (2,500–3,499 RMB per month) in Beijing, 3.68 (1,500–2,499 RMB per month) in Guangzhou, and 2.61 (500–1,499 RMB per month) in Nanchang. The minimum income levels are level I (less than 500 RMB per month) in three cities and the maximum income levels are level X (more than 10,000 RMB per month) in Beijing, level IX (9,000–9,999 RMB per month) in Guangzhou, and level V (3,500–4,499 RMB per month) in Nanchang. Respondents' income levels are spread out in a wide range in Beijing and Guangzhou, while respondents' income levels are centered near the mean value of the subsamples in Nanchang.

It has to be noticed that some workshops were organized at universities. Since students participated in the workshops, most of who earned less than 500 RMB per month, this could explain the great proportion of low income groups. The distribution of income cannot be compared as actual income distribution data are not available.

Table 5.6 Highest educational level distribution

	Beijing ($n=80$)		Guangzhou ($n=80$)		Nanchang ($n=80$)	
	N	(%)	*N*	(%)	*N*	(%)
Illiterate	1	1.3	0	0	0	0
Primary school	0	0	1	1.3	0	0
Junior high school	1	1.3	1	1.3	1	1.3
Senior high school	7	8.8	5	6.3	9	11.3
Bachelor	35	43.8	28	35.0	45	56.3
Master	33	41.3	42	52.5	22	27.5
Doctor	3	3.8	3	3.8	3	3.8

The average monthly income per person in China was 1,315 RMB (China's National Bureau of Statistics 2008). Beijing residents had the highest monthly income, averaging at 2,184 RMB per month. Residents in Guangdong province, the capital of which is Guangzhou, had a monthly income of 1,792 RMB. Income level was about 1,157 RMB per month and per capita in Jiangxi Province where Nanchang is located (China's National Bureau of Statistics 2008). The equality between per capita income in the three cities and the sampling frame can be proved.

5.1.4 Educational Level

Most respondents hold the highest educational level of bachelor (Beijing 43.8%; Guangzhou 35.9%; Nanchang 56.3%) and master (Beijing 41.3%, Guangzhou 52.5%; Nanchang 27.5%) (see Table 5.6). The less educated groups are underrepresented.

The statistics of educational level in China can be found in Table 5.6. The data on educational level in Guangzhou and Nanchang is not available. Therefore, the *Chi*-square test on educational level could not be carried out. Referring to the statistics in Table 5.7, it still can be concluded that the highly educated groups are overrepresented while the lowly educated groups are underrepresented. The sample size is 80 while there are 16.95 million residents in Beijing, 7.73 million residents in Guangzhou, and 4.91 million residents in Nanchang (China's National Bureau of Statistics 2008). It is extremely difficult to guarantee the representativeness of subsamples with various educational levels comparing with the Chinese population.

5.1.5 Family Size

Mean family sizes of the subsamples in Beijing, Guangzhou, and Nanchang are almost equal, 3.43, 3.73, and 3.76, respectively. Nuclear family groups, that is, a

Table 5.7 Statistics of educational level in Beijing, Guangzhou, and Nanchang

City	Illiterate (%)	Primary school (%)	Junior high school (%)	Senior high school (%)	Bachelor, master and doctor (%)
Beijing	3.13	12.62	29.91	22.07	32.27
Guangdong	3.86	27.40	41.02	16.16	11.56
Jiangxi	5.7	33.43	33.66	13.78	13.43

Source: Statistics in China, 2008
Note: The capital of Guangdong Province is Guangzhou and the capital of Jiangxi Province is Nanchang

Table 5.8 Family size distribution

	Beijing ($n=80$)		Guangzhou ($n=80$)		Nanchang ($n=80$)	
	N	(%)	*N*	(%)	*N*	(%)
1	1	1.3	3	3.8	1	1.3
2	7	8.8	7	8.8	3	3.8
3	41	51.3	25	31.3	31	38.8
4	19	23.8	19	23.8	24	30.0
Over 4	12	15.0	26	32.5	21	26.3

Table 5.9 Environmental membership distribution

Are you member of some environmental group(s)?	Beijing ($n=80$)		Guangzhou ($n=80$)		Nanchang ($n=80$)	
	N	(%)	*N*	(%)	*N*	(%)
Yes	21	27.5	5	6.3	25	31.3
No	58	72.5	75	93.8	55	68.8

family unit consisting of mother, father, and their child, account for a great proportion (Beijing 51.3%, Guangzhou 31.3%, Nanchang 38.8%), which is in accordance with the one-child policy in China. Family groups of three, four, and over four are approximately equally represented, albeit there seems to be a bit more participants in nuclear family groups in Beijing. Single groups and family groups of two (dink family) are slightly underrepresented (see Table 5.8). The data on family size distribution in China are not available and, therefore, it cannot be compared between the sampling frame and Chinese population.

5.1.6 Membership of Environmental Groups

Only a small proportion of respondents are member of environmental group(s)/organization(s) (Beijing 27.5%, Guangzhou 6.3%, Nanchang 31.3%) (see Table 5.9). There is no statistical information on how many Chinese join in environmental groups/organizations. Therefore, it cannot be judged whether the proportion of environmental membership is under- or overrepresented.

5.1.7 *Summary*

So far the sociodemographic characteristics of respondents, such as gender proportion, age distribution, income distribution, educational level, family size, and membership of environmental group(s), have been described. Male respondents and female respondents are equally represented in the three subsamples. Average income of respondents is highest in the subsample of Beijing and lowest in the subsample of Nanchang, which is in accordance with the economic situation in the three cities. Most of the respondents are highly educated and received the highest educational degree of bachelor or master. Lower educated respondents are underrepresented. With regard to family size distribution, singles and families of two persons are underrepresented while the larger families (e.g., 3 persons, 4 persons, or more) are well-represented. Finally, a small proportion of individuals hold membership of environmental group(s). Especially in Guangzhou, very few respondents are a member of an environmental organization.

Chi-square analyses were conducted between the subsamples and the population in the three cities on gender distribution and age distribution. Results indicate that all three subsamples are representative in terms of gender distribution and the subsample in Guangzhou is representative in terms of age distribution. Respondents of old age (above 65 years old) are underrepresented in the subsamples of Beijing and Nanchang. Because no official statistics on income distribution, educational level distribution, and family size distribution is available in the three cities, *chi*-square analyses between the sampling frame and the populations in the three cities could not be made. However, it can be estimated that the subsamples are representative in terms of family size, because families of three persons account for the greatest proportion, which is consistent with the one-child policy in China. Low educated individuals (below senior high school) are underrepresented and highly educated individuals (above senior high school) are overrepresented. The income level distribution of individuals in the three subsamples is consistent with the economic situation in the three cities.

In order to test the theoretical validity of WTP responses, respondents were also asked some questions on environment and wildlife protection to examine their environmental attitudes. In the following, I turn to the analysis of respondents' attitude on some general environmental issues and porpoise protection questions.

5.2 Attitudinal Questions

In this section, the goal is to explore respondents' attitude toward the environment and wildlife, their comments on the information folder and the opinion on the porpoise conservation project. The information on respondents' attitudes can help test the validity of WTP bids.

5.2.1 *Attitude on General Issues on Environment and Wildlife*

Six questions on the environment and wildlife issues, which can be found in Appendix D, were put forward to respondents so as to check their attitude on environmental and wildlife issues.

Most respondents (47.5% in Beijing, 51.25% in Guangzhou, 63.75% in Nanchang) watch/listen programs about the environment and nature on TV, Internet, and radio when they can. Very few respondents (1.25% in Beijing, 5% in Guangzhou, 2.5% in Nanchang) hardly ever watch such programs. On the whole, individuals pay attention to the programs about the environment and nature to a great extent. That is, individuals care about the environment and nature basically.

Most respondents (78.75% in Beijing, 81.25% in Guangzhou, 75% in Nanchang) strongly agreed that it is important to take account of the environment when making decisions on economic development. Also respondents (21.25% in Beijing, 18.75% in Guangzhou, 20% in Nanchang) agreed that environmental protection should be considered into the decisions on economic development. Only four respondents in Nanchang did not know whether the environmental protection should be considered when developing economics. Nobody (strongly) disagreed with the importance of considering environment when developing economics.

Correspondingly, most respondents (76.25% in Beijing, 80% in Guangzhou, 71.25% in Nanchang) strongly agreed that human beings should invest in the environment so that the next generation can benefit from the plants and animals on Earth. Many individuals (21.25% in Beijing, 20% in Guangzhou, 26.25% in Nanchang) agreed with the suggestion of investing in the environment for future generations. Very few individuals had no idea whether to invest in the environmental protection for future generations. On the whole, individuals hold positive attitudes on the investment in environment for the sake of sustainable development.

Regarding the protection of endangered rare species, most respondents (88.75% in Beijing, 88.75% in Guangzhou, 87.5% in Nanchang) insist that the number of endangered rare species should be increased. A few respondents (3.75% in Beijing, 3.75% in Guangzhou) think that it is enough to keep the present population. A small number of respondents (6.25% in Beijing, 6.25% in Guangzhou, 7.5% in Nanchang) contended that people should leave endangered species alone. A handful of respondents (1.25% in Beijing, 1.25% in Guangzhou, 5.0% in Nanchang) also told that they had no idea on what to do with these species. In sum, responses in the three cities were similar. It can be concluded that most individuals contend that the preservation of endangered rare species is important and the amount of endangered rare species should be increased.

Most respondents (65% in Beijing, 63.75% in Guangzhou, 60% in Nanchang) strongly agreed that animals have a right to live on Earth even though they might be of no direct use to mankind. Also, many individuals (32.5% in Beijing, 33.75% in Guangzhou, 33.75% in Nanchang) agreed that animals have a right to live on Earth. Several respondents (2.5% in Beijing, 2.5% in Guangzhou, 5.0% in Nanchang) also showed that they did not know how to comment on the protection of animals, especially when they do not have direct use values. Only one respondent in Nanchang disagreed with that all the animals have a right to live on Earth. Basically, individuals respect the existence of animals on the earth, no matter whether they are of direct use to human kinds.

When respondents were asked to decide about policy priority on five issues, including (1) lowering housing price, (2) providing medical and endowment insurance in rural area, (3) solving the graduate unemployment issue, (4) education, and (5) environmental wildlife protection, a significant portion of respondents gave the first priority to lowering housing price (43.75% in Beijing, 45.0% in Guangzhou, 31.25% in Nanchang). Wildlife and environmental protection was given the first priority only by a few respondents (8.75% in Beijing, 2.5% in Guangzhou, 8.75% in Nanchang).

In summary, most individuals paid attention to the environment and wildlife in various ways. They also thought that the environment and nature are very important and agreed that it is necessary to implement some protection measures on environment and rare species. However, when comparing with other issues, such as education, lowering housing prices, and graduate unemployment, the environment and wildlife protection is given the least priority. This ranking of priorities is quite important for later analysis of willingness to pay responses.

5.2.2 Comments on Information Folder

Among the respondents, some of them heard about the YFPs before the survey. The proportion of respondents who have heard about the YFPs is highest in Nanchang, 82.5%, comparing that in Beijing (66.25%) and Guangzhou (68.75%) (see Table 5.10). The closer to Poyang Lake people live, the more opportunities people get to know the YFPs.

After reading and discussing the information folder, respondents could have good knowledge of the YFPs. Many respondents were impressed with their uniqueness. They learned human activities threatened the species. They felt surprised that the YFPs would die out in 25–100 years if no protection measures were taken. They also thought that human beings were mainly responsible for the threats faced by porpoises and should take some measures to protect the YFPs as soon as possible. Overall, the information was considered interesting and easy to understand. At the same time, most participants were happy with the amount of information. A few respondents thought that the information folder was slightly under loaded and stated that they would prefer more information on the biological characteristics of the YFPs, like multiplication rate, biomass structure, distinctive behavioral traits etc. Some questions, like biomass structure, were answered immediately, and other questions, like multiplication rate and longevity, could not be answered due to a lack of relevant scientific research on porpoises by biologists.

Table 5.10 Whether to hear about Yangtze finless porpoises

Have you heard about YFPs	Beijing ($n=80$)		Guangzhou ($n=80$)		Nanchang ($n=80$)	
	N	(%)	*N*	(%)	*N*	(%)
Yes	53	66.25	55	68.75	66	82.5
No	27	33.75	25	31.25	14	17.5

Table 5.11 Respondents' opinion on the conservation project

Do you support the porpoise conservation project?	Beijing ($n=80$)		Guangzhou ($n=80$)		Nanchang ($n=80$)	
	N	(%)	*N*	(%)	*N*	(%)
Yes	78	97.5	78	97.5	75	93.75
No	0	0	1	1.25	2	2.5
Don't know	2	2.5	1	1.25	3	3.75

5.2.3 *Individuals' Views Toward the Porpoise Conservation Project*

Because of the urgent situation that the YFPs face, a nature reserve was proposed in Poyang Lake. Most respondents (97.5% in Beijing, 97.5% in Guangzhou, 93.75% in Nanchang, see Table 5.11) stated that they support the porpoise conservation project based on a series of arguments (Box 5.2).

Box 5.2 Exemplary Statements for Supporting the Porpoise Conservation Project

Beijing:

"Human activities cause the YFPs into a dilemma and it is our duty to compensate them now."

"The project is in accordance with the requirements of sustainable development."

"The YFPs lie top of the food chain and play a key role in maintaining the stability of the whole chain."

"The YFPs are unique in China and are worth conservation."

"They may have great potential values but invisible for the moment."

"The project is a typical example for conservation of rare species."

"The YFPs have a right to life and it is our duty to protect them."

"The habitat of the YFPs was seriously destroyed and it is necessary to protect them from now on."

"Nature reserve is the most effective protection method. More people should participate in the conservation of YFPs."

"The extinction of the YFPs may incur more unknown loss. It is necessary to establish a nature reserve for them."

(continued)

Box 5.2 (continued)

Guangzhou:
"The YFPs have existence values and are important for maintaining the regional ecological balance."

"The YFPs should enjoy the same right to life as human kinds."

"It is our responsibility to protect endangered rare species."

"This project is a good conservation example."

"We should pursue the ecological balance and protect endangered rare species."

"Economic development causes threats to the YFPs and it is time to compensate them."

"The porpoise conservation can promote the environmental protection in Poyang Lake region."

"It is our duty to protect wildlife, including the YFPs."

"River dolphins have died out and we should pay more attention to the YFPs."

"The project is operational and important."

Nanchang:
"The project is a good event for residents in Jiangxi province."

"We should pursue sustainable development and protect endangered rare species, like the YFPs."

"The project can help develop an independent porpoise biomass in Poyang Lake."

"The YFPs plays an important part in gene storage bank."

"I hope my future generations still have the chance to see the YFPs."

"The YFPs are endangered right now. If no protection, we can only see their fossil in a museum."

"The YFPs are mammals, the same as human beings, and we should protect them."

"The conservation project can not only build up a stable 'home' for the YFPs but also improve the environment of Poyang Lake."

"The YFPs lie on top of food web and are very important for maintaining the stability of the whole food web."

"The porpoise population can be increased through the conservation project."

Only a very small part of respondents (2.5% in Beijing, 1.25% in Guangzhou and 3.75% in Nanchang) said that they did not know whether to support the project. Reasons are listed as following:

Beijing: Firstly, I don't know whether the funds for the nature reserve are continuous; secondly, there is no detailed information on public participation in the project. Local residents in Poyang Lake region play a key role on the development of the nature reserve; at last, how can you make sure that local residents' economic loss caused by the development of nature reserve can be compensated? Without these information, I cannot give my judgment.

Beijing: The opinion should be inquired from experts. I failed to judge that.

Guangzhou: I am not sure how the effects of the project will be like. I suspect whether the project can achieve the goal. I don't know.

Nanchang: I don't think that the nature reserve can help the YFPs get rid of risk of extinction.

Nanchang: I prefer Poyang Lake project to the porpoise conservation project. Is there a better idea for the porpoise protection than the nature reserve?

Nanchang: I am confused what is more important, economic development or ecology? Lots of species have died out but our life is still going on. I have no idea whether to support or not.

One individual in Guangzhou and two individuals in Nanchang refused to support the project because of the following reasons:

Guangzhou: The ecosystem of Poyang Lake is integrity and the YFPs are only a part of the ecosystem. People should concentrate on preserving the whole ecosystem, not only one kind of species. Only when the ecology and environment of whole Poyang Lake are improved, can the protection of YFPs be strengthened. The conservation project only aims at the porpoise protection. Will the project produce any side effects on other species? All the information is not certified. I doubt that the project is feasible.

Nanchang: I doubt that the population can be increased only by developing a nature reserve.

Nanchang: I think it is better to leave the YFPs alone. Human's involvement would make the situation worse.

A possible explanation for the statement is they do not intend to support the project or do not know whether to support it or not. Or the participants had difficulties abandoning their previous knowledge on some failure examples of wildlife protection and, therefore had no faith in porpoise conservation project. Nevertheless, most respondents showed their concern on YFPs and would like to support the project.

5.2.4 Individuals' Preference on Different Uses of Porpoise Conservation

Before the WTP question, one question was put forward to respondents that they could explain why they are willing to contribute to the porpoise protection (see Appendix D). As explained in Chap. 3, the TEV of porpoise conservation consists of various types of subvalues, like direct use value, indirect use value, option value, bequest value, and altruistic value. The purpose of the question is to check the validity of WTP results. People's preferences on different values are shown in Fig. 5.1. It can be seen that most respondents (Beijing: 60%; Guangzhou: 63.8%; Nanchang: 57.5%) agreed that the YFPs played an important role in maintaining the biodiversity, which was classified as indirect use value; secondly, a great deal of respondents (Beijing: 67.5%; Guangzhou: 66.3%; Nanchang: 55%) indicated that they want the YFPs to continue to live in the world, which was classified as existence value. In addition, a great proportion of respondents (Beijing: 52.5%; Guangzhou: 47.5%; Nanchang: 42.5%) stated that they hope that future generations could enjoy the existence of the YFPs, which was classified as bequest values of porpoise conservation. On the whole, the indirect use values, bequest values, and existence values of the porpoise protection received most attention from respondents. On average, respondents' preferences on the five kinds of values are almost homogenous among the three valid subsamples.

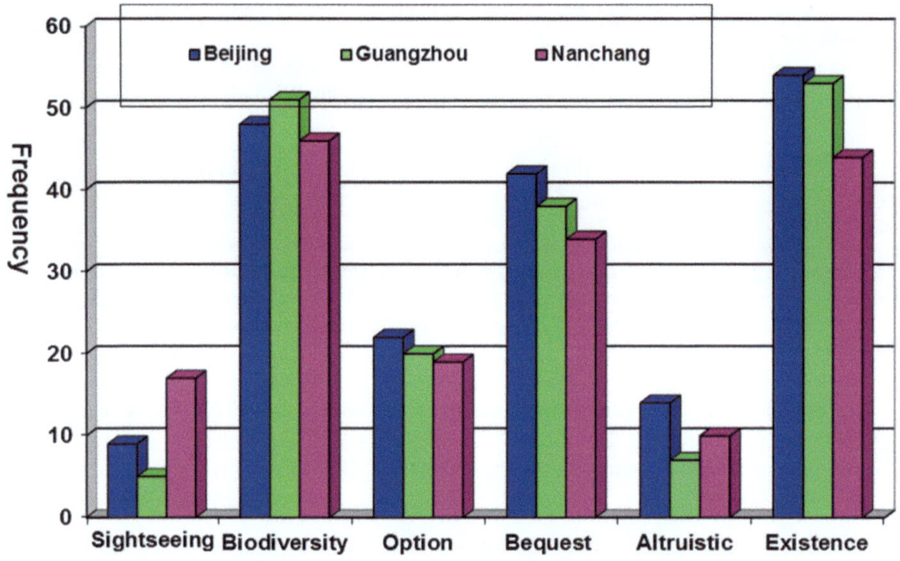

Fig. 5.1 Preference on difference uses of porpoise conservation

When it is already known whether respondents support the project, I can go on to the question whether they are willing to contribute to the project. In the following section, I turn to the willingness to pay and unwillingness to pay responses.

5.3 Validity of Responses

If some respondents do not express their estimated true value, CVM might fail to determine the correct economic value of the good in question. Therefore, it is important to consider the potential for divergences between observed responses generated in the CVM exercise and the true value before analyzing CVM data (Mitchell and Carson 1989). WTP are valid when respondents state the amount that they would be genuinely willing to pay if the environmental change could actually occur (Desvousges et al. 1993). Although assessing the validity of WTP responses of porpoise conservation is complicated by the fact that true WTP is unobservable and hence unknown, some biases such as warm glow, strategic, or protest bidding can be identified. The participants' answers to the following debriefing questions after the WTP question can prove whether their stated bids are valid or not (Hanemann 1994). This section aims to identify protest responses, warm glow, strategic bias, embedding effects, and reports the reasons that participants stated to explain their WTP bids.

5.3.1 Nonrespondents and Nonresponses

Cases and controls who either are not contacted, refused, or were not able to sustain the interview were classified as nonrespondents (Richiardi et al. 2002). Many factors can contribute to nonresponse. Some individuals may fail to answer because of the less interest in the survey object (Mattson and Li 1994). Other individuals may fail to complete and return the questionnaire due to lack of time, reluctance to be involved in any survey, or intellectual limitations to understand the questionnaire and the constructed market scenario (Carson 1991).

Nonrespondents are of three types: unit nonresponse, item nonresponse, and protest nonresponse. I deal with protest responses in the subsequent section (Sect. 5.3.2). Unit nonresponse occurs at both phone interview level and mail level when no information is obtained on the unit of analysis.

Item nonresponse occurs when the individual fails to respond to the contingent market value elicitation question (Whitehead et al. 1993). In this study, nonrespondents only include item nonrespondents to the WTP elicitation question. Given the group-based survey method on porpoise conservation project, item nonresponses refer to those respondents who failed to respond to WTP questions because they were alert on the money issue (e.g., "*I refuse to answer the question because it is about money*.") or do not have faith in that government can effectively manage the conservation project (e.g., "*I will not donate until the conservation project starts.*") In the survey, the item nonresponse rate to WTP question is estimated to be 2.5% in Beijing, 2.5% in Guangzhou, and 1.25% in Nanchang. Babbie (2007) states, "A review of the published social research literature suggests that a response rate of at least 50% is considered adequate for analysis and reporting. A response of 60% is good; a response rate of 70% is very good." Singleton and Straits (2005) note, "It is very important to pay attention to response rates. For interview surveys, a response rate of 85% is minimally adequate; below 70% there is a serious chance of bias." It can, therefore, be inferred that a nonresponse rate of 15–50% is acceptable. On the whole, the nonresponse rates in the study on the YFPs of these three subsamples are low. The response rates are more than sufficient.

5.3.2 Protest Responses

The treatment of protest bids becomes especially important when the benefit aggregation issue is considered (Halstead et al. 1992). It is agreed that some respondents do not state their true bids, because they protest some components of the CV survey (Meyerhoff and Liebe 2006). For example, some respondents may have dissension with specific aspects of the CV study, such as the payment vehicle (Morrison et al. 2000), ethical beliefs indicated, for example, by lexicographic preferences (Söderquist 1998), or fairness aspects (Jorgensen et al. 2001).

Table 5.12 Protest responses of unwilling-to-pay and possible reasons

Reasons of respondents in Beijing	Beijing	Guangzhou	Nanchang
I have no money			•
It is Chinese government's business	•	•	•
Money should be collected from factories around Poyang Lake	•	•	•
Money should be collected from big companies			•
NGOs such as WWF, IUCN, should be responsible for the YFP conservation	•		•
It is not my business		•	
I do not care about porpoises			•
I am afraid that the money will not be used for the conservation project	•	•	•
Other reason:	•	•	•
"My contribution is too small to be of importance." (Beijing)			
"The Chinese government should protect the YFPs using the taxes already paid. If all the rare species protection needs citizens' donation, our life will become difficult."(Beijing)			
"It is not enough to depend only on citizens' donation." (Beijing)			
"Money is not that important for the porpoise protection. The reformation of the whole management system on rare species plays a key role."(Beijing)			
"It is not enough to just concentrate on porpoise conservation. We should work on the protection of the whole environment and ecosystems."(Guangzhou)			
"The method of payment (donation) is considered to be inappropriate." (Nanchang)			
"I am afraid the money will not be used for the porpoise protection. What is worse, the money might be used for illegal purposes by some organizations or groups. There are lots of similar examples in the society."(Nanchang)			
Amount of protest answers (Unwilling to pay)	9	6	11

In CVM studies, some respondents protest by simply refusing to pay any amount of money (i.e., protest responses of unwillingness-to-pay) or by stating untrue bids (i.e., protest responses of willingness-to-pay). In our study on the YFPs, some individuals refused to pay for the increase of the YFP population directly for the following reasons (see Table 5.12).

In sum, there are nine protest responses of unwillingness-to-pay in Beijing, six protest responses of unwillingness-to-pay in Guangzhou, and 11 protest responses of unwillingness-to-pay in Nanchang. The protest response rates of unwillingness-to-pay are 11.25% in Beijing, 7.5% in Guangzhou, and 13.75% in Nanchang (see Table 5.12).

Table 5.13 Approaches to identifying protest answers in CVM

City	Statements
Beijing	"I really do not have to pay extra money."
Beijing	"Such fund should not depend on social donation. More funds should be supported by government and big companies."
Guangzhou	"The amount of donation is not important. What is more important is to inspire residents' awareness on the importance of biodiversity."
Guangzhou	"The amount of donation is not important. It is more important to encourage more people to contribute to the YFP protection."
Guangzhou	"The amount of donation is not important. Tell residents how to do good to environment, like the classification of rubbish."
Guangzhou	"Donation is not that important. More people should be informed about the importance of rare species conservation."
Guangzhou	"Maybe I will donate less because I am worried about the corruption problem."
Guangzhou	"Government and companies should take charge of most expenses."
Nanchang	"100 RMB is only a symbol. It does not mean anything."

In addition, some other respondents protest by stating untrue bids that are called protest responses of willingness-to-pay. This circumstance happens when respondents do not understand what they are asked to do in the survey and they answer the valuation question anyway (Meyerhoff and Liebe 2006). There are no absolute criteria to identify protest bids (Jorgensen et al. 1999). A widely used technique is to add debriefing questions that can, to some extent, distinguish protest bids from true bids (e.g., Hadker et al. 1997; Ryan et al. 2004). This technique was also employed in this study, as the respondents were asked to justify the reasons for rejecting the scenario.

Several protest responses of willingness-to-pay are identified according to respondents' follow-up statements (see Table 5.13).

According to Table 5.13, two *protest responses of willingness-to-pay* are identified from the subsamples in Beijing, six *protest responses of willingness-to-pay* are identified from the subsamples in Guangzhou, and one *protest response of willingness-to-pay* is identified from the subsamples in Nanchang. Then, the *protest response rates of willingness-to-pay* are, respectively, 2.5% in Beijing, 7.5% in Guangzhou, and 1.25% in Nanchang. In a CV study with a deliberative approach of citizens' jury, the protest response rate of willingness-to-pay was up to 10% (Kenyon and Nevin 2001). Therefore, the protest response rates in this study are in all acceptable.

In total, 11 protest responses are identified from the subsamples in Beijing, 13 protest responses from the subsamples in Guangzhou, and 12 protest responses from subsamples in Nanchang, respectively (see Table 5.14). The protest response rates are 13.75% in Beijing, 15% in Guangzhou, and 15% in Nanchang. Protest responses are quite common in CV studies. The percentage of protest responses goes up to 33%, or even 50% in some CV studies (see Halstead et al. 1992; Wronka 2001; González-Cabán et al. 2007). Therefore, the protest response rates are acceptable in this study.

Table 5.14 Protest responses in Beijing, Guangzhou and Nanchang

	Beijing (n=80)		Guangzhou(n=80)		Nanchang(n=80)	
Protest responses	*N*	(%)	*N*	(%)	*N*	(%)
Unwillingness to pay	9	11.25	6	7.5	11	13.75
Willingness to pay	2	2.5	6	7.5	1	2.5
Total	11	13.75	12	15	12	15

It is noteworthy that the majority of the statements are related to the fact that the state was involved in the scenario, a fact which is also emphasized by Meyeroff and Liebe (2006). In the study of Meyerhoff and Liebe (2006), the authors attempt to further investigate the motives of protest bids and conclude that protest bids could be reduced if the scenario did not include elements such as the involvement of the state. However, an alternative hypothetical market without involving the state is not acceptable in the study due to the fact that all such conservation projects in China are constructed and managed by the state.

5.3.3 Strategic Bias

Economists are aware of the existence of systematic incentives for strategic misrepresentation of preferences by consumers in revealing their true demands for public goods. One the one hand, individuals may be motivated to indicate a lower amount because they expect that others' contributions will be sufficient for obtaining the good, which is incurred by the nonexcludable nature of a public good (Mitchell and Carson 1989; Throsby and Withers 1986). This phenomenon is the classic "free rider"' problem. However, research on the phenomenon indicates that in practice free riding is not as prevalent as initially predicted by many economists. Therefore, it does not pose a major threat to the validity of CV studies (Posavac 1998; Cummings et al. 1986). This phenomenon was also not found in our study on porpoise conservation.

On the other hand, if respondents believe bids are purely hypothetical and funds will not be collected, they may overestimate their WTP for an environmental good or service, as this increases the probability that the environmental good or service will be provided and that someone else will have to pay for it (Garrod and Willis 1999). This overbidding problem was also not found in the study on the YFPs.

In addition, some individuals apply biases and heuristics to simplify their decision-making, especially in uncertain and complex conditions (Tversky and Kahneman 1974; Busenitz and Barney 1997; Zajac and Bazerman 1991; Bateman and Zeithaml 1989; Jackson and Dutton 1988). For example, when individuals were asked to value porpoise conservation, some of them applied heuristics to simplify the question and then gave some heuristic WTP bids, which are also classified as strategic bids. They tried to decide how much money would be needed if all or most Chinese

Table 5.15 Strategic reports in the survey

City	Explanation
Beijing	"If everybody donates 10 RMB, 10 billion will be collected in China."
Beijing	"If most citizens would like to donate 50 RMB, lots of money will be collected."
Guangzhou	"I do not have to pay more than 100 RMB. If 1% Chinese would like to donate 100 RMB, 13 million RMB will be collected, which can be enough for the establishment of a nature reserve."
Guangzhou	"I am used to donating 100 RMB every time."
Guangzhou	"10 RMB can be afforded by most people. If everybody would like to donate 10 RMB, lots of money can be collected."
Guangzhou	"If everybody donates 20 RMB, 100 million residents can donate 2 billion, which is enough for building up a nature reserve."
Nanchang	"20 RMB can be afforded by most Chinese. If 10 million people would like to donate 20 RMB, enough funds can be collected for the nature reserve."
Nanchang	"100 RMB is a lucky number."
Nanchang	"50 RMB is the middle number between 0 and 100."
Nanchang	"Everybody should donate. My contribution is not that important."

contribute (see Table 5.15). The proportions of strategic responses are 2.5% in Beijing, 5% in Guangzhou, and 5% in Nanchang. In Table 5.15, some strategic answers of respondents are listed.

5.3.4 Embedding Effect

The embedding phenomenon results from the finding of different WTP values for a good depending on whether the good is valued on its own or as part of a more inclusive package of goods (Kahneman and Knetsch 1992). "Perfect embedding" is said to arise when WTP responses do not vary between questions seeking the value of protecting a small element of the environment (e.g., a species) or a much larger component of the environment (e.g., a landscape) (Bennett et al. 1998). In a number of studies, it has been discovered that respondents are willing to pay (a) the same amount for goods that differ in quality and (b) different amounts for the same good (Carson and Mitchell 1995; Bennett et al. 1998). That is, individuals may offer responses that only appear to ignore differences in levels or amounts of a public good (e.g., Goodman et al. 1998). In addition, a proportion of the embedding effect might arise when respondents offer a zero bid, even though they may well-distinguish higher levels of quality or greater amounts of the public good from lower levels presented in the survey (Jorgensen et al. 2001).

In our study on the YFPs, some respondents gave answers that point at the existence of an embedding effect. The responses are listed in Table 5.16, and it seems they ignored the differences in the levels of a public good. For example, one respondent neglected the difference between the YFPs and all the aquatic animals. Also, four respondents did not pay attention to the difference between the YFP protection

Table 5.16 Embedding effect of responses

City	Statements
Beijing	"My answer reflects my views on the need to preserve all the aquatic animals, not only the YFPs."
Beijing	"The amount of donation is not important, what is more important is whether citizens hold awareness on the wildlife protection."
Beijing	"I would like to contribute to maintaining the harmonious ecology."
Guangzhou	"I am an environmentalist and I would like to contribute to the wildlife protection."
Guangzhou	"My donation reflects on my views on the wildlife protection."
Guangzhou	"I would like to donate 100 RMB for the maintenance of ecological harmony."
Nanchang	"5 RMB represents my duty on the wildlife protection."
Nanchang	"My answer reflects my attention on the environmental protection."

and the whole wildlife protection. Two respondents did not take the differences between porpoise conservation and the maintenance of the ecological status into account when they answered the WTP bid question. One respondent was even confused about the difference between porpoise conservation and environmental protection in general. In sum, the rates of responses with an embedding effect are detected to be 3.75% in Beijing, 3.75% in Guangzhou, and 2.5% in Nanchang.

5.3.5 Warm Glow

Warm glow is a terminology of Adreoni (1989, 1990), which means that individuals get some pleasurable feeling from their giving *per se*, distinct from the benefit enjoyed by the recipient. CV is a stated preference method for estimating the "worth"—in terms of money—that a public good provides for an individual. Based on the private provision of public goods, CV respondents derive moral satisfaction or a warm glow from the giving *per se* (Kahneman and Knetsch 1992). The evidence that the warm glow effect has an important influence on the WTP answers seems to be rather robust (Nunes and Schokkaert 2003).

In our study, only two warm glow responses can be identified in the survey samples of Beijing according to their explanations to the follow-up debriefing questions. The rate of warm glow responses is 2.5% in the subsample of Beijing, rather low. No warm glow effect could be identified in the subsamples of Guangzhou and Nanchang.

> "I am a teacher and I can set a good example for my students. I can get satisfaction from giving to a good cause." (1,000 RMB)
>
> "My answer reflects my views on the need to preserve all the aquatic animals not only the YFPs." (100 RMB)

Table 5.17 Invalid responses distribution

Invalid responses		Beijing ($n=80$) N	Beijing (%)	Guangzhou ($n=80$) N	Guangzhou (%)	Nanchang ($n=80$) N	Nanchang (%)
Nonresponses		2	2.5	2	2.5	1	1.25
Protest responses	Unwilling to pay	9	11.25	6	8.75	11	13.75
	Willing to pay	2	3.75	6	7.5	1	2.5
	Total	11	13.75	12	15	12	15
Strategic bias		2	2.5	4	5	4	5
Embedding effect		3	3.75	3	3.75	2	2.5
Warm glow		2	2.5	0	0	0	0
Sum		20	25	21	26.25	19	23.75

Table 5.18 Invalid, valid, and missing responses distribution

		Beijing ($n=80$) N	Beijing (%)	Guangzhou ($n=80$) N	Guangzhou (%)	Nanchang ($n=80$) N	Nanchang (%)
Valid	WTP	60	75	59	72.5	60	75
	Un-WTP	0	0	0	0	1	1.25
	Sum	60	75	59	73.75	61	76.25
Invalid	WTP	9	12.5	13	16.25	7	8.75
	Un-WTP	9	11.25	6	7.5	11	13.75
	Sum	18	23.75	19	23.75	18	22.5
Missing	WTP	1	1.25	2	2.5	1	1.25
	Un-WTP	1	0	0	0	0	0
	Sum	2	2.5	2	2.5	1	1.25

5.3.6 *Summary of Invalid Responses*

So far all the invalid responses are identified (see Table 5.17). The protest response rates are highest among various invalid responses. All kinds of invalid responses distribute rather evenly among the three subsamples of Beijing, Guangzhou, and Nanchang.

5.3.7 *Valid WTP Responses and Attached Statements*

In total, 60 (75%) valid WTP responses are identified among the subsample of Beijing, 59 (73.75%) in Guangzhou, and 60 (75%) in Nanchang (Table 5.18). In total, the valid WTP response rates in all three cities are higher than 70%. The results can, therefore, support further analysis and calculation of the TEV of the YFPs.

The valid responses are judged according to the follow-up explanations from respondents. Exemplary explanations for WTP are listed in Box 5.3.

Box 5.3 Exemplary Reasons Stated by Participants to Explain Their Valid WTP Bids

Beijing:
"I think the YFPs are important. Threats faced to them are caused by human beings and thus we should compensate them now."
"This is how much I can afford."
"I love the YFPs, the unique rare species in China."
"Donation is only what I can do to porpoise conservation."
"I would like to try my best to contribute to porpoise conservation."
"My answer reflects my views on the need to preserve the genetic material of YFPs."
"The YFPs play an important role in maintaining biodiversity and they are deserved the donation."
"The conservation project can help the YFPs get rid of the danger of dying out and I would like to do them a favor."
"The YFPs are humankinds' good friends."
"It is one-off payment. I would like to do my best to contribute to it."

Guangzhou:
"It is up to my budget constraint."
"It is 1% of my present monthly income."
"To protect the YFPs is the responsibility of the whole society."
"This is my monthly surplus except the necessary expenditure."
"My hometown is in Poyang Lake region. I should contribute to the YFPs."
"The cost of porpoise conservation is great and I should try my best to contribute to it."
"The porpoise protection needs support and sponsorship from many people."
"The porpoise conservation project is of significant important to all the people."
"When I have more money in future, I would like to contribute more."
"My income is not high, but I still want to do my best to help the YFPs."

Nanchang:
"If the collected money can be used for porpoise conservation, everybody will respond actively."
"This is a token of my regard on the YFPs."
"Many hands make light work. I hope more people will participate in the porpoise conservation project."
"It is possible for me to donate more. I am afraid the censorship of the funds is not transparent and money may be used for other purposes."
"I can save such money by buying less clothes and fruits."
"Everybody should try his or her best to help the YFPs."
"If everybody donates a little to the porpoise conservation project, the nature reserve can be built up."
"The YFPs are of great importance."
"It is one-time payment. It is my current maximal affordability."
"I am environmentalist and I love YFPs."

According to the above statements, it can be concluded that many respondents decided their amount of WTP according to their budget constraint. The remaining respondents showed their concern of the YFPs by giving some WTP bids. On the whole, most respondents thought that the YFPs were important and that it would be necessary to build a nature reserve in Poyang Lake to strengthen the porpoise conservation. All the valid WTP bids will be included in the estimation of mean WTP for porpoise conservation.

5.4 Quantitative Analysis

In the previous section, invalid bids were identified and eliminated out of the data sets. The following quantitative analysis investigates the distribution of valid WTP bids, provides the descriptive statistics for data in the three cities, tests the validity of WTP responses statistically, compares the mean WTP in the three cities in order to test the distance effect, and analyses mean WTP in China and Germany. For the sake of comparing mean WTP in the three cities, it is essential to compare some socioeconomic characteristics of individuals giving valid responses in the three cities as a first step.

5.4.1 Comparison of Socioeconomic Characteristics

The analysis of variance (ANOVA) (Box 5.4) is applied to test the differences of some socioeconomic characteristics among the subsamples in Beijing, Guangzhou, and Nanchang.

Box 5.4 Analysis of Variance (ANOVA)

Analysis of variance (ANOVA) can be used to test the null hypothesis (H_0) of equality of several population means. If the test shows significance, we reject H_0: we conclude that there is a difference among the population means.

Basically, the ANOVA works like this: A group mean is taken to be an estimate of people's typical level of performance under that particular condition. But individual performance can vary widely and at times deviates markedly from the group mean. Think of this within group variability as background noise, or error. There is high variability between (i.e., among) groups. The ANOVA F statistic is calculated by dividing an estimate of the variability Between Groups by the variability within groups:

$$F = \frac{\textit{Variance}\text{ between}}{\text{Variance within}}$$

(continued)

Box 5.4 (continued)

If there are large differences among the treatment means, the numerator of F (and, therefore, F itself) will be inflated and the null hypothesis is likely to be rejected; but if there is no effect, the numerator and denominator of F should have similar values, giving an F close to unity. A high value of F, therefore, is evidence against the null hypothesis of equality of all the population means (Kinnear and Gray 2000).

Table 5.19 ANOVA of gender distribution among valid subsamples

	Sum of squares	df	Mean square	*F*	Sig.
Between groups	0.008	2	0.004	0.016	0.984
Within groups	44.729	176	.254		
Total	44.737	178			

Table 5.20 ANOVA of age distribution among valid subsamples

	Sum of squares	df	Mean square	*F*	Sig.
Between groups	0.832	2	0.416	0.259	0.772
Within groups	283.045	176	1.608		
Total	283.877	178			

Five important characteristics are tested in our study: gender proportion, age category, income distribution, educational level, and family size. All test results are listed in the following tables:

1. Gender Distribution (Table 5.19)

$$F(2,176)=0.016;\ p>0.05$$

The degrees of freedom of the numerator are 2, while the degrees of freedom of the denominator are 176. The p-value (Sig.) for the F ratio of 0.016 is listed as 0.984, much larger than 0.05. Therefore, F is statistically insignificant. That is, the null hypothesis H_0 of equality of three population means of gender distribution cannot be rejected. It can be concluded that the gender distributions in the valid subsamples of Beijing, Guangzhou, and Nanchang do not differ from each other.

2. Age Distribution (Table 5.20)

$$F(2,176)=0.259;\ p>0.05$$

The degrees of freedom of the numerator are 2 while the degrees of freedom of denominator are 176. The p-value (Sig.) for the F ratio of 0.259 is listed as 0.772.

Table 5.21 ANOVA of income distribution among valid subsamples

	Sum of squares	df	Mean square	*F*	Sig.
Between groups	56.300	2	28.150	8.508	0.000
Within groups	582.337	176	3.309		
Total	638.637	178			

Table 5.22 Multiple comparisons of income distribution among valid subsamples in China

					95% Confidence interval	
(I) Subsample	(J) Subsample	Mean difference (I–J)	Std. error	Sig.	Lower bound	Upper bound
Beijing	Guangzhou	0.472	0.334	0.335	−0.32	1.26
	Nanchang	1. 350*	0.332	0.000	0.57	2.13
Guangzhou	Beijing	−0.472	0.334	0.335	−1.26	0.32
	Nanchang	0.878*	0.334	0.025	0.09	1.67
Nanchang	Beijing	−1.350*	0.332	0.000	−2.13	−0.57
	Guangzhou	−0.878*	0.334	0.025	−1.67	−0.09

Note: (1) Dependent variable: personal monthly income level
(2) Tukey HSD (Tukey's HSD (Honestly Significant Differences) Test is used for finding the differences between the means of all the .subsamples, The difference score will be compared to a critical value to see if the difference is significant. The critical value in this case is the HSD (honestly significant difference) and it must be computed. It is the point when a mean difference becomes honestly significantly different). $HSD = q\sqrt{\frac{MS_{within}}{n}}$ q is a table value; The Mean Square value is computed from the ANOVA; n is the number of the values dealt with in each subsample (Howell 1997).
*The mean difference is significant at the 0.05 level

When the p-value of F is larger than 0.05, F is statistically insignificant. Then the null hypothesis H_0 of equality of the three population means of age category can be accepted. Valid subsamples in three cities are similar in terms of age distribution.

3. Income Distribution (Table 5.21)

$$F(2,176) = 8.508; P < 0.01$$

The degrees of freedom of the numerator are 2 while the degrees of freedom of denominator are 176. The p-value (Sig.) for the F ratio of 8.508 is listed as 0.00. When the p-value of F is much less than 0.01, F is statistically significant. Then the null hypothesis H_0 of equality of the three population means of income level can be rejected.

Inspection of p-values (see Table 5.22) shows that Nanchang valid subsample differs significantly ($p < 0.05$) from both Beijing and Guangzhou valid subsamples in terms of income level at the significant level of 0.05, while Beijing valid subsample does not differ from Guangzhou valid subsample in this case.

Table 5.23 ANOVA of educational level distribution among valid samples

	Sum of squares	df	Mean square	*F*	Sig.
Between groups	1.442	2	0.721	1.093	0.337
Within groups	116.111	176	0.660		
Total	117.553	178			

Table 5.24 ANOVA of family size distribution among valid samples

	Sum of squares	df	Mean square	*F*	Sig.
Between groups	3.344	2	1.672	1.667	0.192
Within groups	176.578	176	1.003		
Total	179.922	178			

Checking the original data of income level of respondents who offered valid responses in Nanchang, the average income level is lower than those in Beijing and Guangzhou, which is in accordance with the actual economic situation in these three cities. If more respondents of higher income were included in the valid subsample of Nanchang, the results could have been improved.

4. Educational Level Distribution (Table 5.23)

$$F(2,176) = 1.093; p > 0.05$$

The degrees of freedom of the numerator are 2, while the degrees of freedom of denominator are 176. The *p*-value (Sig.) for the *F* ratio of 1.093 is listed as 0.337. When the *p*-value of *F* is much higher than 0.05, *F* is statistically insignificant. Then the null hypothesis H_0 of equality of three population means of educational level can be accepted. That is, mean educational levels of respondents distribute equally among the valid subsamples of three cities.

5. Family Size Distribution (Table 5.24)

$$F(2,176) = 1.667; p > 0.05$$

The degrees of freedom of the numerator are 2, while the degrees of freedom of denominator are 176. The *p*-value of (Sig.) for the *F* ratio of 1.667 is listed as 0.192. When the *p*-value of *F* is much bigger than 0.05, *F* is statistically insignificant. Then the null hypothesis H_0 of equality of the three population means of family size can be accepted. That is, mean family size distributes equivalently among the valid subsamples of the three cities.

6. Summary

In sum, valid subsamples do not differ statistically from each other in Beijing, Guangzhou, and Nanchang in terms of gender proportion, age category, educational level, and family size. If the mean income level of individuals in Nanchang were a bit higher, the results could have been better.

5.4.2 *Descriptive Statistics*

A number of statistical descriptors can be sued to evaluate the mean WTP for the conservation of YFPs.

The mean, otherwise called the arithmetic mean or average, is the most commonly used measure of central tendency. The estimation of mean value of WTP bids is a cardinal measure of the welfare benefits that may derive from porpoise conservation. In order to calculate the mean, I simply have to add up all the valid WTP bids in the group of data and divide them by the number of corresponding participants in a workshop.

The median is described as the numeric value separating the higher half of a sample from the lower half. The median WTP can be found by arranging all the observations from lowest WTP value to highest WTP value and picking the middle one. If there is an even number of observations, then there is no single middle value, so one often takes the mean of the two middle values.

The standard deviation of a data set is the square root of its variance. Standard deviation shows how much variation there is from the "average" (mean). It may be thought of as the average difference of the WTP bids distribution from the mean WTP, how far they are away from the mean WTP. A low standard deviation indicates that the WTP bids tend to be very close to the mean WTP, whereas high standard deviation indicates that the WTP bids are spread out over a large range of WTP values.

The standard error of mean is used to refer to an estimate of that standard deviation, derived from a particular sample used to compute the estimate. The standard error of mean is usually estimated by the sample estimate of the population standard deviation divided by the square root of the sample size (assuming statistical independence of the values in the sample).

The confidence interval is a particular kind of interval estimate of a population parameter and is used to indicate the reliability of an estimate. How likely the interval is to contain the parameter is determined by the confidence level or confidence coefficient. Increasing the desired confidence level will widen the confidence interval. The most used confidence coefficient is 95%.

According to the statistics, 70 individuals (87.5%) would like to pay for the porpoise conservation project in Beijing, 73 (91.3%) agree to pay for the project in Guangzhou, and 68 (85%) are willing to pay for that in Nanchang (see Table 5.25).

The high percentage of rejecting the scenario (e.g., 54%, Halvorsen 1996) and refusing to pay any amount of money is not rare in CV studies (e.g., 65%, Dziegielewska and Mendelsohn 2005). The percentages of willing-to-pay in the subsamples of the three cities are high enough for making valid WTP analysis in the study.

Based on the individuals' explanation on their WTP bids, some biased responses, such as protest responses, strategic behavior, embedding effects, and warm glows have been identified, which has been elucidated in Sect. 5.3. CVM is based on the

Table 5.25 Unwillingness or willingness to pay for the porpoise conservation project

Are you willing to pay for the project?	Beijing ($n=80$)		Guangzhou ($n=80$)		Nanchang ($n=80$)	
	N	(%)	*N*	(%)	*N*	(%)
Yes	70	87.5	73	91.3	68	85.0
No	9	11.3	6	7.5	12	15.0
Missing	1	1.3	1	1.3	0	0

assumption that respondents are able to construct unique economic values for these goods by referring to their own well-constructed and known preference orderings (Chilton and Hutchinson 2000). When respondents failed to express their true preferences on the YFP protection project, their WTP bids were removed for any further analysis.

Comparing the Valid WTP Means in Beijing, Guangzhou, and Nanchang

In order to test the distance effect on WTP, valid WTP means in the three cities are compared with the ANOVA analysis. Then, a regression analysis and a correlation analysis are conducted to check the influence of the variable *Distance* on WTP to be significant or not. First, whole valid WTP responses of the three subsamples are compared. Secondly, valid WTP responses of the same income level groups are compared.

1. *Whole valid WTP responses*
   According to Table 5.26, it can be found that mean WTP of the Beijing subsample is largest while that of the Guangzhou subsample is smallest. Furthermore, ANOVA is applied to analyze the difference of valid WTP means in Beijing, Guangzhou, and Nanchang (see Table 5.27). The degrees of the numerator of the F ratio (here, it is the Between Group df) are 2, while the degrees of the denominator (here, it is the Within Groups df) of the F ratio are 170. The p-value (Sig.) for the F ratio of 2.270 is listed as 0.106. The p-value of F is larger than 0.05, F is statistically insignificant.[4] That is, the null hypothesis H_0 of equality of valid WTP means in the three cities cannot be rejected.

$$F(2,170)=2.270;\ p>0.05$$

[4] In statistics, a result is called "statistically significant" if the probability of its occurrence by chance is less than 5%. Chance refers to unexpected, unplanned, unpredictable event that occurs without observable cause or human intention, and is not explainable by the known laws of science or statistics (http://www.businessdictionary.com).

Table 5.26 Descriptive statistics for WTP estimates in Beijing, Guangzhou, and Nanchang

		Beijing	Guangzhou	Nanchang
N		60	59	60
Mean		90.93	54.08	61.02
Median		100.00	50.00	50.00
Standard deviation		137.480	41.744	67.248
Standard error of mean		17.749	5.435	8.682
Minimum bid		1	1	1
Maximum bid		1,000	200	500
95% Confidence interval of the difference	Lower	55.42	43.21	43.64
	Upper	126.45	64.96	78.39

Note: (1) Currency: RMB
(2) 1 RMB = 0.109385 Euro (April, 2010)

Table 5.27 ANOVA of valid WTPs in Beijing, Guangzhou, and Nanchang

	Sum of squares	df	Mean square	*F*	Sig.
Between groups	6,457.975	2	3,228.987	2.270	0.106
Within groups	241,773.991	170	1,422.200		
Total	248,231.965	172			

Here, it can be concluded that there is no evidence for the heterogeneity of variances among valid WTP means of the three subsamples. That is, valid WTP means of the three subsamples do not differ from one another statistically.

Comparing WTP means of the three subsamples serve for testing the distance effects on WTP. As has been explained in Chap. 2, distance refers to the distance between respondent's residence and Poyang Lake. The a priori expectation of the mean WTP was that the mean WTP in Beijing < mean WTP in Guangzhou < mean WTP in Nanchang due to distance effects. However, the results indicate that the mean WTP in Beijing > mean WTP in Nanchang > mean WTP in Guangzhou. The mean WTP in Nanchang is bigger than that in Guangzhou, which is consistent with the theoretical prediction of distance effects, while the mean WTP in Beijing is biggest among the three subsamples, which contradicts with the theory of distance decay.

First of all, the correlation between the variable *Distance* and the WTP of the subsamples in Guangzhou and Nanchang was tested to testify the distance effects. The coefficients are negative, showing that WTP decreases with increasing distance. However, p-values were larger than 0.05 (1-tailed), proving WTP is not highly correlated with distance in the subsamples of Nanchang and Guangzhou (see Table 5.28). That is, distance does not play a significant role in determining the WTP in the two subsamples of Nanchang and Guangzhou.

$$\tau = -0.032;\ n = 119;\ p > 0.05$$

Table 5.28 Correlation between the variable distance and WTP in the subsamples of Nanchang and Guangzhou

		Distance	WTP
Distance	Kendall's tau_b	1.000	−0.032
	Sig. (1-tailed)	0.000	0.346
	N	119	117
WTP	Kendall's tau_b	−0.032	1.000
	Sig. (1-tailed)	0.346	.
	N	119	119

Table 5.29 Correlation between the variable distance and WTP in Beijing, Guangzhou, and Nanchang

		Distance	WTP
Distance	Kendall's tau_b	1	0.120*
	Sig. (1-tailed)		0.027
	N	179	179
WTP	Kendall's tau_b	.120*	1
	Sig. (1-tailed)	0.027	
	N	179	179

*Correlation is significant at the 0.05 level (1-tailed)

Secondly, the correlation between the variable *Distance* and the WTP in the three subsamples was tested with SPSS. The results suggest that the variable *Distance* is highly correlated with WTP at the 0.05 level (1-tailed). The coefficients are positive, indicating that WTP increases with increasing distance (see Table 5.29).

$$\tau = 0.120; n = 179; \ p < 0.05$$

2. Valid WTP responses of the same income level groups
   WTP responses were also compared within the same income level groups. The results are listed in Table 5.30. The mean WTP is always highest in the subsample of Beijing. Among the low income groups (less than 1,500 RMB per month), the mean WTP in Guangzhou is higher than that in Nanchang. Among the high income groups (1,500–4,499 RMB per month), the mean WTP is higher in Nanchang than that in Guangzhou.
   ANOVA analyses were also made to test the difference of WTP means among the same income groups (see Table 5.31). Because the *p*-values are both bigger than 0.05, the mean WTP bids do not differ statistically among the low income groups, so neither among the high income groups.
   When checking the correlation between *Distance* and the WTP bids, it is discovered that *Distance* is positively but not highly correlated with WTP in the low income groups, and also it is not closely related with WTP bids in the high income groups (see Table 5.32). That is, WTP bids of the low income groups of the high income groups do not change with distance, respectively.

Table 5.30 WTP bids of different income groups in Beijing, Guangzhou, and Nanchang

City	Mean	Std. deviation	Min.	Max.	Std. error of mean	Median	*N*
Low income group of less than 1,500 RMB per month							
Beijing	58.33	39.716	5	100	10.255	60.00	15
Guangzhou	48.21	39.557	1	100	9.075	40.00	19
Nanchang	36.67	24.436	5	100	4.988	36.00	24
High income group of 1,500 RMB–4499 RMB per month							
Beijing	113.58	184.809	1	1,000	33.193	71.76	31
Guangzhou	49.82	37.527	5	100	7.092	35.71	28
Nanchang	77.25	80.953	1	500	13.492	75.00	36

Table 5.31 ANOVA analysis of WTP bids of low and high income groups in Beijing, Guangzhou, and Nanchang

Low income group (less than 1,500 RMB per month)			(less than 1,500 RMB per month)		
	Sum of squares	df	Mean square	*F*	Sig.
Between groups	4,465.020	2	2232.510	1.919	0.156
Within groups	63,981.825	55	1163.306		
Total	68,446.845	57			
$F(2,55) = 1.919, p > 0.05$					
High income group (1,500–4,499 RMB per month)					
	Sum of squares	df	Mean square	*F*	Sig.
Between groups	60,631.552	2	30,315.776	2.159	0.121
Within groups	1,292,026.406	92	14,043.765		
Total	1,1352,657.958	94			
$F(2,92) = 2.159, p > 0.05$					

Table 5.32 Correlation between variable distance and WTP in Beijing, Guangzhou, and Nanchang

Low income group (less than 1,500 RMB per month)		WTP	Distance
Distance	Kendall's tau_b	1.000	0.155
	Sig. (1-tailed)		0.081
	N	58	58
WTP	Kendall's tau_b	0.155	1.000
	Sig. (1-tailed)	0.081	0.000
	N	58	58
$\tau = 0.155$; $n = 58$; $p > 0.05$			
High income group (1,500–4,499 RMB per month)		WTP	Distance
Distance	Kendall's tau_b	1.000	0.068
	Sig. (1-tailed)		0.190
	N	95	95
WTP	Kendall's tau_b	0.068	1.000
	Sig. (1-tailed)	0.190	0.000
	N	95	95
$\tau = 0.068$; $n = 95$; $p > 0.05$			

In summary, three mean WTP bids in the three subsamples do not differ statistically significantly on the whole, for low income groups (less than 1,500 RMB per month) or for high income groups (1,500–4,499 RMB per month). (1) On the whole, the mean WTP of the Beijing subsample is biggest and the mean WTP of the Guangzhou subsample is lowest (mean WTP in Beijing > mean WTP in Nanchang > mean WTP in Guangzhou). WTP bids are significantly and positively correlated with the variable distance at the 0.05 level. (2) For the low income groups (less than 1,500 RMB per month), mean WTP in Beijing is biggest while mean WTP in Nanchang is lowest (mean WTP in Beijing > mean WTP in Guangzhou > mean WTP in Nanchang). WTP is not highly correlated with distance. (3) For the high income groups (1,500–4,499 RMB per month), mean WTP in Beijing is highest while mean WTP in Guangzhou is lowest (mean WTP in Beijing > mean WTP in Nanchang > mean WTP in Guangzhou). WTP is not highly correlated with distance. After comparing mean WTP in the subsamples of Beijing, Guangzhou, and Nanchang, I go on to compare mean WTP between China and Germany.

Comparing WTP Means in China and Germany

1. Whole valid responses in China and Germany

   On the whole, WTP for the conservation of YFPs in Germany is much higher than that in China. The mean WTP in German cities Leipzig and Halle is 18.33 Euro, about twice as much as in Beijing (9.94 Euro), three times as that in Guangzhou (5.92 Euro), and two and half times as that in Nanchang (6.67 Euro) (see Table 5.33). Similarly, maximum WTP in Germany is also larger than that in China. Median WTP in Germany (10 Euro) almost equaled that in Beijing (10.94 Euro) and is about twice as those in Guangzhou (5.47 Euro) and Nanchang (Euro). It has to be noticed that there are four zero WTP bids among the valid samples of Germany while the minimum WTP bids are 0.11 Euro (=1 RMB) among the valid samples in China.

   ANOVA analyses were conducted among the subsamples in China and Germany. The results are listed in Table 5.34. The degrees of numerator of F ratio (here, it is the Between Group df) are 3, while the degrees of denominator (here, it is the Within Groups df) of F ratio are 229. The p-value (Sig.) for the F ratio of 7.936 is listed as 0.00. That is, the p-value of F is much smaller than 0.01, and F is statistically significant. Thus the null hypothesis of equality of valid WTP means in the three cities can be rejected. There is strong evidence for the heterogeneity of variances among valid WTP means of the subsamples of China and Germany.

$$F(3,229) = 7.936;\ p < 0.01$$

   Inspection of the p-values (Sig.) for the mean difference between Leipzig-Halle and Guangzhou, Beijing, Nanchang are 0.000, 0.017, and 0.000, respectively (see Table 5.35), much smaller than 0.05, showing that the subsample of Germany differs significantly from the three subsamples of China statistically in terms of

Table 5.33 Descriptive statistics of WTP estimates in China and Germany (in Euro)

Site		Beijing	Guangzhou	Nanchang	Halle-Leipzig
N		60	59	60	54
Mean WTP		9.94	5.92	6.67	18.33
Std. deviation		15.04117	4.56648	7.36083	25.37418
Std. error		1.94181	0.59450	0.95028	3.45299
Median WTP		10.94	5.47	5.47	10
95% Confidence interval for mean	Lower bound	6.0611	4.7301	4.7680	11.4033
	Upper bound	13.8322	7.1102	8.5710	25.2549
Minimum WTP		0.11	0.11	0.11	0
Maximum WTP		109.40	21.88	54.69	125

Note: (1) 1 CNY = 0.109385 Euro (April, 2010)
(2) Data of Halle-Leipzig are cited from Herkle (2010)

Table 5.34 ANOVA of WTP estimated in China and Germany

	Sum of squares	df	Mean square	*F*	Sig.
Between groups	5,393.772	3	1,797.924	7.936	0.000
Within groups	51,878.139	229	226.542		
Total	57,271.911	232			

Table 5.35 Multiple comparisons of WTP estimates among valid subsamples in China and Germany

(I) Group	(J) Group	Mean Difference (I–J)	Std. Error	Sig.	95% Confidence interval Lower bound	Upper bound
Leipzig-Halle	Guangzhou	12.40890*	2.83459	0.000	5.0730	19.7448
	Beijing	8.38241*	2.82328	0.017	1.0758	15.6890
	Nanchang	11.65957*	2.82328	0.000	4.3529	18.9662

Note: *The mean difference is significant at the 0.05 level
(1) Dependent variable: individual WTP bids
(2) Tukey HSD

mean WTP, while the three subsamples of China are homogeneous subsets, which is already stated in Sect. 5.4.2.1. In short, mean WTP in Germany is statistically larger than those in China.

2. Valid responses of the same income groups (below 500 Euro per month) in China and Germany
   Furthermore, WTP of respondents of low income level (below 500 Euro per month) are also compared (see Table 5.36). Among the samples at low income level, the mean WTP is highest in Beijing, followed by Halle-Leipzig. The lowest mean WTP is in Guangzhou.
   ANOVA analyses are also conducted among the subsamples of low income groups in China and Germany. The results are listed in Table 5.37. The degrees of numerator of *F* ratio (here, it is the Between Group df) are 3, while the degrees of denominator (here, it is the Within Groups df) of *F* ratio are 171. The *p*-value

Table 5.36 WTP estimates of respondents at the low income level (below 500 Euro per month) in China and Germany (in Euro)

Site		Beijing (China)	Guangzhou (China)	Nanchang (China)	Halle-Leipzig (Germany)
N		51	51	60	13
Mean WTP		9.81	5.4331	6.67	8.59
Std. deviation		16.22234	4.10636	7.36083	6.62817
Std. error		2.27158	0.57501	0.95028	1.83832
Median WTP		6.56	5.47	5.47	6.25
95% Confidence interval for mean	Lower bound	5.2517	4.2782	4.7680	4.5908
	Upper bound	14.3769	6.5881	8.5710	12.6015
Minimum WTP		0.11	0.11	0.11	1
Maximum WTP		109.40	10.94	54.69	20

Table 5.37 ANOVA of WTP estimated by low-income respondents (below 500 Euro per month) in China and Germany

	Sum of squares	df	Mean square	*F*	Sig.
Between groups	545.018	3	181.673	1.753	0.158
Within groups	17,725.252	171	103.656		
Total	18,270.270	174			

(Sig.) for the *F* ratio of 1.753 is listed as 0.158. That is, the *p*-value of *F* is much larger than 0.05, and *F* is statistically insignificant. The null hypothesis of equality of valid WTP means in the three cities can be accepted. There is strong evidence for the homogeneity of variances among valid WTP means of the low income subsamples of China and Germany. It can be concluded that low income respondents' WTP bids do not differ from each other in Germany and China statistically, while the mean WTP in Germany is a bit higher than those in Guangzhou and Nanchang of China.

5.4.2.3 Summary

In summary, the mean WTP in Beijing is higher than that in Guangzhou, while the mean WTP in Guangzhou is lower than that in Nanchang. The result is not consistent with the theoretical prediction of distance effects. However, the difference of the mean WTP estimates of the three subsamples in China is not statistically significant. The mean WTP estimate in Germany is larger than those in China and the difference of the mean WTP between China and Germany is statistically significant. However, when comparing WTP bids of respondents at the low income level of less than 500 Euro per month, the mean WTP is highest in Beijing, followed by Leipzig-Halle, and lowest in Guangzhou. However, the mean WTP bids of the low-income subsamples do not differ statistically. In order to understand the different WTP estimates, it is necessary to find out what factors play a major role in determining the WTP by checking the theoretical validity of WTP. This will be done in the subsequent section.

5.4.3 Theoretical Validity of WTP

One objective of the analysis is to assess the validity of WTP responses quantitatively to demonstrate that WTP responses are not purely random. Here, I concentrate on testing the theoretical validity, the most feasible way for validity assessment in CV studies, which has been explained in Chap. 4. A multiple linear regression analysis (Box 5.5) is run in SPSS to test the estimates of WTP against the factors that determine WTP.

Box 5.5 Multiple Regression

In multiple regression, the values of the dependent variable Y is estimated from those of two or more independent variables ($x_1, x_2, \cdots, x_p$). This is achieved by the construction of a linear equation of the general form

$$y' = b_0 + b_1(x_1) + b_2(x_2) + \cdots + b_p(x_p),$$

where the parameters $b_1, b_2 \cdots, b_p$ are the partial regression coefficients and the intercept is the regression constant. This equation is known as the multiple linear regression equation of y upon $x_1, x_2, \cdots, x_p$.

In a multiple linear regression WTP bids are regressed against a series of independent variables that are likely to influence WTP bids. The output of regression answers the following the questions (Backhaus et al. 1996, pp. 17–31):

How well does the equation fit to the data?
Which independent variables have a significant influence on WTP?
How does the independent variable influence WTP?

It is a positively biased estimate of the proportion of the variance of the dependent variable accounted for by regression. Mitchell and Carson (1989, p. 213) suggest that a minimum value of 0.15 is needed to show the robustness of the linear relationship between mean WTP and variables.

The regression ANOVA is intended to test whether there really is a linear relationship between the variables by forming an F ratio of the mean square for regression to the residual mean square. When p-value is less than 0.05, F value is significant (Kinnear and Gray 2000).

The Beta weight shows the change in the dependent variable (expressed in standard deviation units) that would be produced by a positive increment of one standard deviation on the independent variable concerned. The independent variable with the largest standardized Beta weights also has the largest correlation with the dependent variable (Kinnear and Gray 2000).

The t-test is to test the regression coefficient for significance. When the p-value of t is less than 0.05, t is significant well-beyond the 0.05 level for the corresponding variable. The signs of t indicate the direction of the influence of independent variable WTP on dependent variable WTP (Backhaus et al. 2000; Kinnear and Gray 2000).

Table 5.38 Independent variables of the regression analysis

Independent variable	Coding
Socio-economic characteristics	
Envi.Group	1=Yes
(Membership in an environmental group)	2=No
AGE	1=less than 16 up to
(Age category)	9=over 85
Edu.Degree	1=illiterate up to
(Highest educational level one receives)	7=doctor
Fam.Size	1=1 person up to
(Amount of members in one household)	5=over 4 persons
INCOME	1=less than 500 up to
(Personal income RMB per month)	10=over 10,000
Environmental attitudes	
ATTENTION	1=make a special effort to watch/listen down to
(Attention on programs about environment and nature on TV/Internet/radio)	4=hardly ever watch/listen to them
Envi.Pro.Priority.	1=most important down to
(Priority of environment and wildlife protection)	5=least important
Econ.AndEnvi.	1=strongly agree down to
(Consider environment when developing economy)	5=strongly disagree
Envi.ForFuture	1=strongly agree down to
(Invest in environment for the benefits of future generation)	5=strongly disagree
AnimalPro.	1=strongly agree down to
(Animals have a right to live on earth)	5=strongly disagree
RareSpe.Pro.	1=agree
The population of endangered rare species should be increased	2=disagree
Overviews on the porpoise protection	
SIGHTSEEING	1=Agree
I have seen the YFPs and I think they are worth protecting.	2=Disagree
BIODIVERSITY	1=Agree
The YFPs play an important role in maintaining biodiversity.	2=Disagree
FutureOption	1=Agree
I would go to Poyang Lake for the sightseeing of the YFPs if possible in the future.	2=Disagree
BEQUEST	1=Agree
I hope future generations can enjoy the existence of the YFPs.	2=Disagree
ALTRUISTIC	1=Agree
I hope some other people can enjoy the existence of the YFPs.	2=Disagree
EXISTENCE	1=Agree
I want the YFPs to continue live in the world.	2=Disagree

In a first step, all the possible factors are explained that can theoretically influence respondents' decision on WTP amount. Basically, there are three kinds of independent variables, namely (1) socioeconomic characteristics, (2) the opinion on general environmental and natural issues, and (3) overviews on the protection of YFPs. All independent variables are listed in Table 5.38.

Table 5.39 Parameter estimates for the WTP bids in Beijing

	Unstandardized coefficients		Standardized coefficients		
	B	Std. error	Beta	*t*	Sig.
(Constant)	64.544	32.306		1.998.	0.052
Edu.Degree	10.609	4.864	0.249	2.181	0.035
INCOME	5.616	2.123	0.301	2.645	0.011
BEQUEST	−19.728	9.506	−0.245	−2.075	0.044
Econ.AndEnvi.	−39.749	12.029	−0.388	−3.304	0.002

In a second step, iterative methods were applied to optimize the model parameters. The number of observations made on each variable should be significantly greater than the number of variables. Otherwise, serious over-fitting may be caused in multivariate linear regression models (Miller 1995; Broadhurst et al. 1997). Because the valid subsample sizes are, respectively, 59, 58, and 60 in Beijing, Guangzhou, and Nanchang, four or less variables are recommended for the regression analysis against WTP. It has to be noticed, some variables may be correlated; in this case, replicated information is redundant (Broadhurst et al. 1997). The signal ratio to noise ratio of certain variables may be so low that their inclusion in the model may be questioned and will certainly lead to a poorer model, especially if other "cleaner" correlated variables are available (Seasholtz and Kowalski 1993; Broadhurst et al. 1997). Therefore, it is necessary to check the correlation between variables before conducting the multiple linear regression analysis. A stepwise multiple regression was applied to pick out an optimal subset among the variables. During the stepwise regression, some predictors were removed subsequently because they do not contribute significantly to the regression. Finally, the reliable predictors were chosen for the dependent WTP in Beijing, Guangzhou, and Nanchang. The multiple regression equations of WTP were developed for the three subsamples, respectively.

5.4.3.1 Identification of Important Variables for WTP Bids in Beijing

In Beijing, four reliable predictors, *Edu.Degree, INCOME*, *BEQUEST*, and *Econ.AndEnvi.* are finally chosen for the multiple regression equation of the dependent variable WTP (see Tables 5.39 and 5.40). The *R* square is 0.434 (see Table 5.41), and hence 43.4% of the variance of the WTP in Beijing can be explained by regression (5.1).

$$\begin{aligned} DependentWTP_{Beijing} &= 10.609 \times (Edu.Degree) + 5.616 \\ &\quad \times (INCOME) - 19.728 \times (BEQUEST) \\ &\quad - 39.749(Econ.AndEnvi.) + 64.544. \end{aligned} \tag{5.1}$$

Table 5.40 Regression ANOVA of variables in Beijing

Model	Sum of squares	df	Mean square	F	Sig.
Regression	33,392.123	4	8348.031	8.423	0.000
Residual	43,609.877	44	991.134		
Total	77,002.000	48			

Table 5.41 The values of multiple R and R^2 in Beijing

R	R^2	Adjusted R^2	Std. error of the estimate
0.659	0.434	0.382	31.482

The result of the regression ANOVA (Table 5.44) informs that there is a significant[5] relationship between the dependent variable and the three independent variables at the 0.5% level.

The Beta weights in Table 5.39 tell that the variable *Econ.AndEnvi.* exerts the strongest influence on WTP, while the variable *BEQUEST* plays the least significant role in the equation. The variable *INCOME* plays the second significant role in determining WTP bids in Beijing, followed by the variable *Edu.Degree*.

The t-values in Table 5.39 indicate that people who receive higher educational level would like to contribute more to the porpoise protection project than those who receive lower educational level. Equivalently, rich people are likely to pay more for the project than poor people. It is also implied that people who want to protect the YFPs for future generations tend to pay more than those who do not want. In the end, the individuals who strongly agree with the statement that environment should be taken into account when developing the economy, pay more than those who do not agree to this statement.

In the study, the goal is to understand how the various variables impact WTP bids. Therefore the multicollinearity is a big problem because individual p-value can be misleading (e.g., a p-value can be very high, even though the variable is important) and the confidence intervals on the regression coefficients will be so wide that excluding a variable or adding a new one can change the coefficients dramatically and may even change their signs. The best solution is to remove one of the two related variables, which does not seem logically essential to the model (Motulsky 1995–2002).

In our study, for example, the variables *AGE*, *RareSpe.Pro.*, *Envi.Pro.Priority, and ATTENTION* are removed from the equation becase they are highly correlated with the significant variable *Edu.Degree*. Similarly, the variables *Fam.Size*, *Envi.ForFuture*, *BIODIVERSITY* are deleted from the model because they are highly correlated with the variable *BEQUEST*. Likewise, the variable *AnimalPro.*is eliminated since it is highly related with the variable *Econ.AndEnvi.*

[5] Statistics. of or pertaining to observations that are unlikely to occur by chance and that therefore indicate a systematic cause (http://dictionary.reference.com/browse/significantly).

In addition, the variables *Envi.Group, FutureOption, ALTRUISTIC, EXISTENCE,* and *SIGHTSEEING* do not enter the model because they do not play a significant role in determining WTP in the Beijing subsample. The survey results show that a small proportion of individuals would like to conserve the YFPs for their present (12%) or possible future (30%) porpoise sightseeing. However, only 20% of the respondents would like to contribute for other people who are interested in the protection of YFPs. It is possible that other individuals who care about such functions of the YFPs still value porpoise conservation less than those who do not care. Most respondents (75%) answered that they would like to conserve the YFPs so that porpoises could continue to live on Earth, some of whom would like to pay less than those who did not mention the point for some reasons, such as budget constraints. Therefore, the above five variables do not exert significant influence on deciding WTP and are eliminated in the model.

Identification of Important Variables for WTP Bids in Guangzhou

Among the valid samples of Guangzhou, four reliable predictors, *Envi.Group, SIGHTSEEING, BIODIVERSITY*, and *INCOME,* were finally selected for the multiple regression equation of dependent variable WTP (see Table 5.42). The R^2 is 0.434 (Table 5.44), hence 43.4% of the variance of the WTP bids in Guangzhou can be explained by regression (5.2).

$$\begin{aligned} DependentWTP_{Guangzhou} = & -42.255(Envi.Group) - 53.409 \\ & \times (SIGHTSEEING) - 28.023(BIODICERSITY) \\ & + 5.437(INCOME) + 250.359 \end{aligned} \quad (5.2)$$

The result of the regression ANOVA can be seen in Table 5.43. It can be proved at the 0.00 level that there is a significant relationship between the dependent variable WTP and the four independent variables Envi.*Group, SIGHTSEEING, BIODIVERSITY*, and INCOME (Table 5.44).

The Beta weights (see Table 5.42) indicate that the variable *Envi.Group* produces most influence on individual WTP in Guangzhou followed by the variable *SIGHTSEEING*. The variable *BIODIVERSITY* takes the third position. The variable *Income* exerts significant but least influence on individual WTP in the Guangzhou subsample, comparing the other three variables.

The *t*-values (see Table 5.42) show that respondents with environmental membership offer more to the porpoise conservation project than those without environmental membership. Likewise, people with higher income are likely to pay more than those with lower income. Equally, people who pay attention to the sightseeing and the ecological function of the YFPs in maintaining biodiversity would like to contribute more than those who do not care.

Some common variables such as *AGE, Edu.Degree, and Fam.Size* are not selected for the regression model against WTP in Guangzhou for the following reasons: The variable *AGE* and variable *Fam.Size* are excluded to prevent the impact of

Table 5.42 Parameter estimates for the dependent WTP in Guangzhou

	Unstandardized coefficients		Standardized coefficients		
	B	Std. error	Beta	t	Sig.
(Constant)	250.359	40.196		6.228	0.000
Envi.Group	−42.255	15.306	−0.303	−2.761	0.008
SIGHTSEEING	−53.409	15.396	−0.383	−3.469	0.001
BIODIVERSITY	−28.023	8.643	−0.356	−3.242	0.002
INCOME	5.437	2.118	.280	2.567	0.013

Table 5.43 Regression ANOVA of variables in Guangzhou

Model	Sum of squares	df	Mean square	F	Sig.
Regression	31,217.239	4	7804.310	9.380	0.000
Residual	40,769.298	49	832.026		
Total	71,986.537	53			

Table 5.44 The values of multiple R and R^2 in Guangzhou

R	R^2	Adjusted R^2	Std. error of the estimate
0.659	0.434	0.387	28.845

multicollinearity, because they are both highly correlated with the variable *INCOME*. Here, the independent variable *Edu.Degree* is poorly correlated with dependent WTP in Guangzhou. It has to be noticed that some students participated in the workshops in Guangzhou. The students have gained the highest educational degree (bachelor or master) and did not offer higher WTP than those with lower educational degree because of their economic conditions. Such kinds of respondents take a great proportion among the valid samples and weaken the linear relationship between the variable *Edu.Degree* and the dependent WTP.

In addition, the variable *Envi.Pro.Priority* is excluded because it is closely related with the variable *INCOME*. Likewise, the variable *ALTRUISTIC* is eliminated from the model because of its close correlation with the variable *BIODIVERSITY*. The remaining variables, *ATTENTION, Envi.ForFuture, AnimalPro., RareSpe.Pro., Econ.AndEnvi., BEQUEST, FutureOption, and EXISTENCE,* are not included in the regression model because they are not correlated with individual WTP bids. These eight variables do not play a role in determining WTP in the subsample of Guangzhou.

Identification of Important Variables for WTP Bids in Nanchang

In the valid samples of Nanchang, the three reliable predictors, *Income*, *Fam.Size*, and *Envi.Group* are finally decided as determinants for the multiple linear regression equation of dependent variable WTP (see Tables 5.45 and 5.46). The R^2 is 0.440 (see Table 5.47), and hence 44.0% of the variance in WTP bids in Nanchang can be explained by regression (5.3).

Table 5.45 Parameters estimate for dependent WTP in Nanchang

Dependent WTP in Nanchang	Unstandardized coefficients		Standardized coefficients		
	B	Std. error	Beta	*t*	Sig.
(Constant)	118.491	23.845		4.969	0.000
Envi.Group	−17.597	7.882	−0.251	−2.233	0.031
Fam.Size	−16.199	4.022	−.460	−4.027	0.000
INCOME	9.765	2.971	0.373	3.286	0.002

Table 5.46 Regression ANOVA of variables in Nanchang

Model	Sum of squares	df	Mean square	*F*	Sig.
Regression	23,963.730	3	7987.910	11.780	0.000
Residual	30,513.821	45	678.085		
Total	54,477.551	48			

Table 5.47 Values of multiple *R* and R^2 in Nanchang

R	R^2	Adjusted R^2	Std. error of the estimate
0.663	0.440	0.403	26.040

$$Dependent\,WTP_{Nanchang} = 9.765 \times (INCOME) - 16.199(\text{Fam.Size}) - 17.597(Envi.Group) + 118.491 \quad (5.3)$$

The result of the regression ANOVA shows that there is strong evidence at the 0.00 level that there is a significant relationship between the dependent variable WTP and the independent variables *Income*, *FamilySize*, and *Envi.Group* (see Table 5.45).

The Beta weights in Table 5.45 state that the variable *Fam.Size* exerts most significant influence on WTP in Nanchang, followed by variables *INCOME* and *Envi.Group*. The *t*-values (see Table 5.45) indicate that people with higher income have higher WTP bids than people with lower income. Equally, WTP decreases with the increase of respondents' family size. Similarly, respondents with environmental membership are willing to contribute more than those without environmental membership. All these results are in line with theoretical expectations.

Since the variable *INCOME* was selected in the model, the variables *AGE*, *ATTENTION*, *Envi.Pro.Priority*, *SIGHTSEEING*, and *ALTRUISTIC* were excluded to avoid the impact of multicollinearity. Similarly, the variables *Edu.Degree, FutureOption, and BIODIVERSITY* were not selected as they were highly correlated with variable the *Envi.Group*. The variables *Econ.AndEnvi*, *AnimalPro*, and *Envi.ForFuture* are highly correlated with each other and none of them play a significant role in determining WTP. The variables *RareSpe.Pro.* and *EXISTENCE* are both highly correlated with the variable *BEQUEST.* None of them is highly correlated with the dependent variable WTP—therefore, they do not enter the regression model.

Table 5.48 Parameters estimate for dependent WTP in Nanchang and Guangzhou

	Unstandardized coefficients		Standardized coefficients		
Model	B	Std. error	Beta	*t*	Sig.
(Constant)	44.910	29.437		1.526	0.130
ATTENTION	−11.644	4.440	−0.229	−2.622	0.010
Envi.Group	−18.738	7.486	−.219	−2.503	0.014
Edu.Degree	8.715	4.053	0.187	2.150	0.034
INCOME	6.984	1.954	0.310	3.574	0.001

Table 5.49 Regression ANOVA of variables in Guangzhou and Nanchang

Model	Sum of squares	df	Mean square	*F*	Sig.
Regression	34,234.006	4	8558.501	9.158	0.000[a]
Residual	92,520.831	99	934.554		
Total	126,754.837	103			

Table 5.50 Values of multiple R and R^2 in Guangzhou and Nanchang

R	R^2	Adjusted R^2	Std. error of the estimate
0.520	0.270	0.241	30.570

Identification of Important Variables for WTP Bids in Beijing, Guangzhou and Nanchang

1. WTP bids in Guangzhou and Nanchang
   Mean WTP in Guangzhou is smaller than mean WTP in Nanchang, which is consistent with the theory of distance effects. However, the variable *DISTANCE* is not correlated with dependent WTP of the two subsamples in Guangzhou and Nanchang. Specifically, the variable *DISTANCE* does play a role in determining individual WTP in Guangzhou and Nanchang. Multiple linear regression analysis was made against WTP in Guangzhou and Nanchang. The results indicate that there is a significant linear relationship between dependent WTP of Guangzhou and Nanchang and the four significant determinants, *ATTENTION*, *Envi.Group*, *Edu.Degree*, and *INCOME* at the 0.00 level ($R^2=0.27$) (see Tables 5.48, 5.49, and 5.50).

   Beta weights in Table 5.48 suggest that the variable *INCOME* exerts the greatest influence on deciding WTP in Guangzhou and Nanchang. The variables *ATTENTION* and *Envi.Group* play the second and third significant roles, respectively, in determining WTP. The variable *Edu.Degree* plays the least significant role.
2. WTP bids in Beijing, Guangzhou, and Nanchang
   Mean WTP in Beijing is biggest among the three subsamples, which contradicts with the theory of distance effect. The variable *DISTANCE* is positively highly correlated with dependent WTP in three cities, which means WTP increases with increased distance. A regression was run to find out whether the variable *Distance* exerts significant influence on respondents' WTP in the three subsamples (see Table 5.51).

Table 5.51 Regression of dependent WTP in Beijing, Guangzhou, and Nanchang against distance

Model	Unstandardized coefficients B	Unstandardized coefficients Std. error	Standardized coefficients Beta	*t*	Sig.
(Constant)	41.016	7.283		5.631	0.000
DISTANCE	7.658	3.383	0.175	2.263	0.025

R: 0.175; R^2: 0.031; Adjusted R^2: 0.025; *N*: 164; *F*: 5.123; *Sig.*:0.025

Table 5.52 Correlation between distance and income in Beijing, Guangzhou, and Nanchang

		Distance	Income
Distance	Kendall tau_b	1	0.216**
	Sig. (1-tailed)		0.000
	N	179	179
Income	Kendall tau_b	0.216**	1
	Sig. (1-tailed)	0.000	
	N	179	179

**Correlation is significant at the 0.01 level (1-tailed)

When WTP was regressed against the variable *Distance* alone, *Distance* works as a significant variable in the regression equation at the condition $p<0.05$ (see Table 5.51). The *t*-statistic of variable *Distance* equals 2.263, which means that WTP increases with increasing distance from respondents' residence to the Poyang Lake. R^2 is less than 0.15. Therefore, the linearity relationship between *WTP* and *Distance* is deemed unreliable**.**

Multiple linear regression analyses are made against dependent WTP in the three cities. Because the variable *DISTANCE* is highly correlated with the variable *INCOME* at the 0.00 level (1-tailed) (see Table 5.52) and distance does not seem to be a logical variable in the equation. The variable *DISTANCE* is excluded from the regression equation to eliminate the impact of multicollinearity.

Three significant determinants, *Envi.Group*, *Edu.Degree* and *INCOME* (see Table 5.53), were selected in the equation. 31.7% variance of WTP estimates in the three cities can be explained by the three variables at the 0.00 level. Beta weights in Table 5.53 indicate that variable *INCOME* plays the most significant role in determining WTP in the three subsamples. The variable *Envi.Group* plays the second significant role in determining WTP, and the variable *Edu.Degree* exerts the least significant influence (Tables 5.54 and 5.55).

3. WTP bids of the high income groups

   Mean WTP is highest in Beijing for both low income group (less than 1,500 RMB per month) and high income group (1,500–4,499 RMB per month). At the same time, mean WTP of the low income groups is bigger in Guangzhou than in Nanchang, while mean WTP of the high income groups is smaller in Guangzhou than in Nanchang. Multiple linear regression analysis was made against WTP of high income groups. The results indicate that there is a significant

Table 5.53 Parameters estimate for dependent WTP in Beijing, Guangzhou, and Nanchang

Model	Unstandardized coefficients B	Unstandardized coefficients Std. error	Standardized coefficients Beta	*t*	Sig.
(Constant)	25.886	20.435		1.267	0.207
Envi.Group	−27.489	5.967	−0.312	−4.607	0.000
Edu.Degree	9.491	3.131	0.205	3.031	0.003
INCOME	8.260	1.337	0.418	6.177	0.000

Table 5.54 Regression ANOVA of variables in Beijing, Guangzhou, and Nanchang

Model	Sum of squares	df	Mean square	*F*	Sig.
Regression	68,015.365	3	22,671.788	23.182	0.000
Residual	146,697.291	150	977.982		
Total	214,712.656	153			

Table 5.55 Values of multiple R and R^2 in Beijing, Guangzhou, and Nanchang

R	R^2	Adjusted R^2	Std. error of the estimate
0.563[a]	0.317	0.303	31.273

Table 5.56 Parameters estimate for dependent WTP bids of high income groups (1,500–4,499 RMB per month) in Beijing, Guangzhou, and Nanchang

	Unstandardized coefficients B	Unstandardized coefficients Std. error	Standardized coefficients Beta	*t*	Sig.
(Constant)	51.069	32.579		1.568	0.121
Envi.Group	−17.877	9.147	−0.198	−1.954	0.054
Edu.Degree	7.601	4.273	0.181	1.779	0.079
BIODIVERSITY	−12.844	7.835	−0.168	−1.639	0.105
FutureOption	12.272	8.425	0.149	1.457	0.149

R: 0.400; R^2: 0.160; Adjusted R^2: 0.121; N: 92; F: 4.145; Sig.: 0.004

linear relationship between dependent WTP of high income group and four significant determinants, *Envi.Group*, *Edu.Degree, BIODIVERSITY, and FutureOption* at the 0.01 level (R^2=0.160) (see Table 5.56). However, no variable works as a predictor for the WTP bids of high income groups.

4. WTP bids of low income groups

   No reliable linear regression model can be developed between dependent WTP of low income groups and any variables. Only the variables *Edu.Degree* is found to be highly correlated with WTP bids of low income groups at 0.05 level (1-tailed) (see Table 5.57).

Table 5.57 Correlation relationship among educational degree and WTP of low income group (<1,500 RMB per month) in Beijing, Guangzhou, and Nanchang

		WTP	Edu.Degree
WTP	Kendall tau_b	1.000	0.205*
	Sig. (1-tailed)		0.035
	N		
Edu.Degree	Kendall tau_b	0.205*	1.000
	Sig. (1-tailed)	0.035	
	N	58	58

*Correlation is significant at the 0.05 level (1-tailed)

Table 5.58 Summary of regression results

Independent variable	WTP in Beijing	WTP in Guangzhou	WTP in Nanchang
Significance	0.000	0.000	0.000
R^2	0.434	0.434	0.440
Determinates	Edu.Degree, INCOME, BEQUEST, Econ. AndEnvi.	Envi.Group, SIGHTSEEING, BIODIVERSITY, INCOME	Income, Fam.Size, Envi. Group

Summary

From Table 5.58, it can be found that the variable *INCOME* is always a reliable predictor throughout all the three regression runs. The variable *Envi.Group* is also significant in influencing individual WTP in Guangzhou and Nanchang. Some common socioeconomic variables like *AGE*, *Fam.Size*, *Edu.Degree*, and *Envi.Group* are excluded from the regression equations due to the high correlations with other significant variables. Here, I focus on the variables of environmental attitudes and the variables of overviews on the porpoise conservation project.

Some environmental attitude variables, like *Econ.AndEnvi*, *AnimalPro*, *Envi.ForFuture*, and *RareSpe.Pro.* are not included in the regression equations. Respondents' answers on the attitude questions are very positive and active. Differences between respondents' attitudes are that between strongly agree and agree, which is too small to explain different WTP bids. Over the past years, the environment in China degraded strongly and people are indeed suffering a lot from environmental pollution and disasters. Snow storm in 2009 spring festival and earthquake in 2008 give all Chinese a serious warning. Also, there are lots of reports on environmental disasters on TV/Radio/Internet. Most Chinese come to realize that environmental protection is important. However, the relationship between WTP and these variables are very insignificant. That is, the variables exert weak influence on WTP.

According to the regression results, the explanatory power is good for WTP estimates in Beijing, Guangzhou, and Nanchang. There is evidence for the significant relationship between the dependent variable WTP and some independent variables at 0.0% levels. All multiple regression coefficients (R^2) are also above 0.4, which means that over 40% of the variance of the WTP estimates can be explained by the regression equations. All Beta weights provide evidence in support of the proposition that WTP estimates measure the intended construct on the basis of socioeconomic characteristics, which is very important for proving the validity of WTP estimates.

5.5 Conclusion

In this chapter, the sociodemographic characteristics of respondents, such as gender proportion, age distribution, income distribution, educational level, family size, and membership of environmental group(s), are described and compared with the official population statistics. The three subsamples in Beijing, Guangzhou, and Nanchang are representative in terms of gender proportion. If a few more old individuals (above 65) and low educated respondents (lower than bachelor degree) were recruited to the workshops, the survey results could have been improved.

Furthermore, some invalid responses, such as nonresponses to WTP question, protest responses, strategic behavior, embedding effect, and warm glow are identified based on individuals' statements on WTP and Un-WTP. The valid response rates of the subsamples are 75, 73.75, and 75% in Beijing, Guangzhou, and Nanchang respectively, which can support further analysis of WTP estimates.

Moreover, socioeconomic characteristics of valid subsamples are compared and the results indicate that the valid subsamples do not differ statistically from each other in Beijing, Guangzhou, and Nanchang in terms of gender proportion, age category, educational level, and family size. If mean income level of individuals in Nanchang were a bit higher, the three valid subsamples would be equivalent statistically in terms of income distribution.

In addition, the mean WTP estimates are compared among the three cities. The result is that mean WTP in Beijing > mean WTP in Nanchang > mean WTP in Guangzhou while they do not differ from each other statistically. The mean WTP results are not consistent with the theoretical prediction of distance effects of mean WTP in Beijing < mean WTP in Guangzhou < mean WTP in Nanchang. Furthermore, mean WTP bids of the same income groups are also compared. Among the low income groups (less than 1,500 RMB per month), mean WTP in Beijing > mean WTP in Guangzhou > mean WTP in Nanchang. Among the high income groups (1,500–4,499 RMB per month), mean WTP in Beijing > mean WTP in Nanchang > mean WTP in Guangzhou.

Additionally, the theoretical validity of valid individual WTP estimates is assessed. It is found out that dependent WTP bids of the respondents in Beijing have a significant relationship with the determinants *Edu.Degree, INCOME*, *BEQUEST,* and *Econ.AndEnvi.* at the 0.00 level (R^2=0.434). Dependent WTP estimates in Guangzhou are significantly related with the variables *Envi.Group, SIGHTSEEING, BIODIVERSITY*, and *INCOME* at the 0.00 level (R^2=0.434). The 44% variance of WTP bids in Nanchang can be well-explained by the variables *Income*, *Fam.Size*, and *Envi.Group* at the 0.00 level.

Furthermore, mean WTP estimates are compared between China and Germany. The mean WTP in Germany is significantly larger than that in China. Mean WTP responses of the same income groups (less than 500 Euro per month) are also compared. The result is that mean WTP in Beijing > mean WTP in Leipzig-Halle > mean WTP in Nanchang > mean WTP in Guangzhou. It should be clear that the mean WTP bids of the same income groups do not differ statistically.

References

Adreoni, J. (1989). Giving with impure altruism: Applications to charity and Ricardian equivalence. *Journal of Political Economy, 97*, 1447–1458.

Adreoni, J. (1990). Impure altruism and donations to public goods: A theory of warm-glow giving. *The Economic Journal, 100*, 464–477.

Alberini, A., and Flores, N. (1995). A bibliography of contingent valuation studies and papers. Natural Resource Damage Assessment, Inc., CA. San Diego. 186pp.

Babbie, E. (2007). *The practice of social research* (11th ed.). Belmont: Wadsworth.

Backhaus, K., Erichson, B., Plinke, W., & Weiber, R. (1996). *Multivariate Analysemethoden*, Berlin:Springer.

Backhaus, K., Erichson, B., Plinke, W., & Weiber, R. (2000). *Multivariate analysemethoden*. Berlin/Heidelberg/New York: Springer.

Bateman, T. S., & Zeithaml, C. P. (1989). The psychological context of strategic decisions: A model and convergent experimental findings. *Strategic Management Journal, 10*, 59–74.

Bennett, J., Morrison, M., & Blamey, R. (1998). Testing the validity of responses to contingent valuation questioning. *The Australian Journal of Agricultural and Resource Economics, 42*, 131–148.

Broadhurst, D., Goodacre, R., Jones, A., Rowland, J. J., & Kel, D. B. (1997). Genetic algorithms as a method for variable selection in multiple linear regression and partial least squares regression, with applications to pyrolysis mass spectrometry. *Analytica Chimica Acta, 348*, 71–86.

Busenitz, L. W., & Barney, J. W. (1997). Differences between entrepreneurs and managers in large organizations: Biases and heuristics in strategic decision-making. *Journal of Business Venturing, 12*, 19–30.

Carson, R. T. (1991). Constructed markets. In J. B. Braden & C. D. Kolstad (Eds.), *Measuring the demand for environmental quality* (pp. 121–162). Amsterdam: North-Holland.

Carson, R. T., & Mitchell, R. C. (1995). Sequencing and nesting in contingent valuation surveys. *Journal of Environmental Economics and Management, 28*, 155–173.

Carson, R. T., Wright, J. L., Carson, N., Alberini, A., & Flores, N. (1995). A bibliography of contingent valuation studies and papers. *Natural Resource Damage Assessment*, Inc., CA. San Diego. 186pp.

Chilton, S. M., & Hutchinson, W. G. (2000). A note on the warm glow of giving and scope sensitivity in contingent valuation studies. *Journal of Economic Psychology, 21*, 343–349.

China Popin. (2008). [Online] Available: http://www.cpirc.org.cn/tjsj/tjsj_cy_detail.asp?id=9229. 13 Mar 2008.

China's National Bureau of Statistics. (2008). Available: http://www.stats.gov.cn/english/statisticaldata/yearlydata/

China State Statistics Bureau. (2008). *China population and employment statistics yearbook 2008*. Beijing: China Statistics Press.

Cummings, R. G., Brookshire, D., & Schulz, W. D. (1986). *Valuing environmental goods: An assessment of the contingent valuation methodology*. Savage: Rowman & Littlefield Publishers.

Desvousges, W. H., Johnson, F. R., Dunford, R. W., Boyle, K. J., Hudson, S. P., & Wilson, N. (1993). Measuring natural resource damages with contingent valuation: Tests of validity and reliability. In J. A. Hausman (Ed.), *Contingent valuation: A critical assessment* (pp. 91–159). Amsterdam: North Holland.

Dziegielewska, D. A. P., & Mendelsohn, R. (2005). Valuing air quality in Poland. *Environmental and Resource Economics, 30*, 131–163.

Edwards, S. F., & Anderson, G. D. (1987). Overlooked biases in contingent valuation surveys: some considerations. *Land Economics, 63*(2), 168–178.

Garrod, G., & Willis, K. G. (1999). *Economic valuation of the environment: Methods and case studies*. Cheltenham: Edward Elgar Publishing.

González-Cabán, A., Loomis, J. B., Rodriguez, A., & Hesseln, H. (2007). A comparison of CVM survey response rates, protests and willingness-to-pay of native Americans and general population for fuels reduction policies. *Journal of Forest Economics, 13*, 49–71.

Goodman, S. L., Seakbrooke, W., & Jaffry, S. A. (1998). Considering conservation value in economic appraisal of coastal resources. *Journal of Environmental Planning and Management, 41*, 313–336.

Hadker, N., Sharma, S., David, A., & Muraleedharan, T. R. (1997). Willingness-to-pay for Borivli National Park: Evidence from a contingent valuation. *Ecological Economics, 21*, 105–122.

Halstead, J. M., Luloff, A. E., & Stevens, T. H. (1992). Protest bidders in contingent valuation. *Northeastern Journal of Agriculture Resource Economics, 21*, 160–169.

Halvorsen, B. (1996). Ordering effects in contingent valuation surveys: Willingness to pay for reduced health damages from air pollution. *Environmental and Resource Economics, 8*, 485–499.

Hanemann, M. (1994). Valuing the environment through contingent valuation. *Journal of Economic Perspectives, 8*, 19–43.

Herkle, S. (2010). Eine ökonomische Bewertung des Jangtse Glattschweinswals mit der Kontingenten Bewertungsmethode, Master Thesis, Martin-Luther University of Halle-Wittenberg.

Howell, D. C. (1997). *Statistical methods for psychology* (4th ed.). Boston: Duxbury.

Jackson, S. E., & Dutton, J. E. (1988). Discerning threats and opportunities. *Administrative Science Quarterly, 33*, 370–387.

Jorgensen, B. S., Syme, G. J., Bishop, B. J., & Nancarrow, B. E. (1999). Protest responses in contingent valuation. *Environmental and Resource Economics, 14*, 131–150.

Jorgensen, B. S., Wilson, M. A., & Heberlein, T. A. (2001). Fairness in the contingent valuation of environmental goods: Attitude toward paying for environmental improvements at two levels of scope. *Ecological Economics, 36*, 133–148.

Kahneman, D., & Knetsch, J. L. (1992). Valuing public goods: The purchase of moral satisfaction. *Journal of Environmental Economics and Management, 22*, 57–70.

Kenyon, W., & Nevin, C. (2001). The use of economic and participatory approaches to assess forest development: A case study in the Ettrick Valley. *Forest Policy and Economics, 3*, 69–80.

Kinnear, P. R., & Gray, C. D. (2000). *SPSS for windows made simple release 10*. Hove: Psychology Press Ltd.

Levy, P. S., & Lemeshow, S. (1999). *Sampling of populations: Methods and applications* (3rd ed.). New York: Wiley-Interscience.

Loomis, J., & King, M. (1994). Comparison of mail and telephone-mail contingent valuation surveys. *Journal of Environmental Management, 41*, 309–324.

Mattsson, L., & Li, C. Z. (1994). Sample nonresponse in a mail contingent valuation survey: An empirical test of the effect on value inference. *Journal of Leisure Research, 26*, 182–188.

Meyerhoff, J., & Liebe, U. (2006). Protest beliefs in contingent valuation: Explaining their motivation. *Ecological Economics, 57*, 583–594.

Miller, J. (1995). *Multiple regression development with GDCC*. Master thesis, University of Montana.

Mitchell, R. C., & Carson, R. T. (1989). *Using surveys to value public goods: The contingent valuation method*. Baltimore: Johns Hopkins Press.

Morrisson, M. D., Blamey, R. K., & Bennett, J. W. (2000). Minimising payment vehicle bias in contingent valuation studies. *Environmental and Resource Economics, 16*, 407–422.

Motulsky, H. (1995–2002). *Multicollinearity in multiple regression*. Retrieved from http://www.graphpad.com/articles/Multicollinearity.htm

National Bureau of Statistics People's Republic of China. (2009). *China population statistics 2009*. Beijing: China Population and Development Research Center. Retrieved from http://www.cpdrc.org.cn

Nunes, P. A. L. D., & Schokkaert, E. (2003). Identifying the warm glow effect in contingent valuation. *Journal of Environmental Economics and Management, 45*, 231–245.

Posavac, S. S. (1998). Strategic overbidding in contingent valuation: Stated economic value of public goods varies according to consumers' expectations of funding source. *Journal of Economic Psychology, 19*, 205–214.

Richiardi, L., Boffetta, P., & Merletti, F. (2002). Analysis of nonresponse bias in a population-based case-control study on lung cancer. *Journal of Clinical Epidemiology, 55*, 1033–1040.

Ryan, M., Scott, D. A., & Donaldson, C. (2004). Valuing health care using willingness to pay: A comparison of the payment card and dichotomous choice methods. *Journal of Health Economics, 23*, 237–258.

Seasholtz, M. B., & Kowalski, B. (1993). The parsimony principle applied to multivariate calibration. *Analytica Chimica Acta, 277*, 165–177.

Singleton, R. A., & Straits, B. C. (2005). *Approaches to social research* (4th ed.). New York: Oxford University Press.

The State Council of People's Republic of China (2001). China's 4th census, retrieved from http://www.stats.gov.cn/tjgb/rkpcgb/.

Söderquist, T. (1998). Why give up money for the Baltic Sea. *Environmental and Resource Economics, 12*, 249–254.

Throsby, C. D., & Withers, G. A. (1986). Strategic bias and demand for public goods: Theory and an application to the arts. *Journal of Public Economics, 31*, 307–327.

Tversky, A., & Kahneman, D. (1974). Judgment under uncertainty: Heuristics and biases. *Science, 185*, 1124–1131.

Whitehead, J. C., Groothuis, P., & Blomquist, G. C. (1993). Testing for non-response and sample selection bias in contingent valuation: Analysis of a combination phone/mail survey. *Economics Letters, 41*, 215–220.

Wronka, T. C. (2001). Protestantworten – theoretischer Hintergrund, empirischer Befund und Lösungsstrategien. In P. Elsasser & J. Meyerhoff (Eds.), *Ökonomische Bewertung von Umweltgütern. Anwendung der Kontingenten Bewertung im deutschsprachigen Raum* (pp. 161–183). Marburg: Metropolis.

Zajac, E. J., & Bazerman, M. H. (1991). Blind spots in industry and competitor analysis: Implications for interfirm (Mis)perceptions for strategic decisions. *Academy of Management Review, 16*, 37–56.

Chapter 6
Discussion

Abstract The research aim of this study is to estimate the values of porpoise conservation by eliciting individuals' WTP using CVM. CVM is rather novel in China and most people are not familiar with paying for environmental goods or services. At the same time, many people do not have much knowledge of the YFPs. Hence, difficulties with the application of CVM are not beyond expectation. Nevertheless, it is still proved that CVM study has been successfully implemented. Facing the difficulty and exploring the reasons is necessary for improving future research.

In this study, the mean WTP in Beijing, Guangzhou, and Nanchang of China is compared in order to test the distance effects on the WTP for porpoise conservation, which is important to establish the market extent of porpoise conservation. When the total values of porpoise conservation are estimated, it is a key step to settle the market extent and to decide whose WTP should be taken into consideration when aggregating the values. Contrary to expectations, the WTP for porpoise conservation in China was not influenced by distance decay.

Furthermore, this study also compared the mean WTP between China and Germany. Until now, there are very few international comparisons of CVM studies. Most CVM studies concentrate on investigating people's preferences for a certain environmental good in a local region or nationwide. YFPs are genetically unique in the world, living in the Yangtze River. It is very meaningful to know people's preference for them outside China, because their conservation is not only important for Chinese people, but also for people all over the world. Germany is taken as an example. The final results indicate that the mean WTP in Germany is much higher than the mean WTP in China.

Keywords Mean WTP • Contingent valuation method • Workshop • Protest response • Distance effect

Y. Dong, *Contingent Valuation of Yangtze Finless Porpoises in Poyang Lake, China*,
DOI 10.1007/978-94-007-2765-6_6,

In this chapter, I seek to discuss the main results. At the beginning, the difficulty with the implementation of CVM in China is discussed, from the recruitment of respondents to the implementation of valuation workshops to those individuals giving biased responses (Sect. 6.1). Afterward, I explain why no distance decay on the WTP occurred in China (Sect. 6.2). Finally, the reasons for differences in WTP in China and Germany are explored (Sect. 6.3). The chapter ends with a brief summary (Sect. 6.4).

6.1 Difficulties with the Design of Contingent Valuation Studies in China

There have been relatively few CVM studies in China until now. Only two CVM studies are concerned with endangered rare species—black-faced spoonbills and giant pandas. In China, decision-making regarding the WTP for conserving endangered rare species is challenging for some individuals. On the one hand, people are not familiar with the concept of paying for public goods. On the other hand, people may not feel responsible for environmental protection because they are used to the fact that the government takes decisions. Therefore, it was not accidental that difficulties and problems during the implementation process of this CVM study emerged, from the recruitment to group discussions.

6.1.1 Recruitment and Implementation of Workshops

The recruitment took place on streets and on university campuses. People always walked in a hurry and only very few people stopped and were patient enough to listen to the explanation of the project. Then, they judged whether they are interested in the topic and decided whether they would like to attend the workshops.

A number of factors let people decide against participating in the valuation workshops: the time they need to spend on the way to and from and during the valuation workshop is long. Usually, the workshops took place from 7 to 9 pm. Three to four hours were demanded from every participant in total, including travel time and the workshop itself. The long time made some individuals hesitate to join the workshops; at the same time, working overtime is wide-spread in China. For example, some individuals' employers unexpectedly required overtime work so that they had to cancel attending the workshops; after 1 day's work, people are exhausted and are not motivated to attend a workshop requiring a lot of brainwork; the reimbursement of 50 RMB (about 5 Euros) did not seem very attractive for some citizens.

The above factors increased the difficulties of recruitment and influenced the components of respondents. Hence, it is likely that the sample contained many participants that are favorably inclined to environmental issues. However, these factors are common and unavoidable in China. In addition to these above factors, there are other reasons affecting the selection of samples. In this section, I focus on explaining why the distribution of age, educational level, and family size in the sample is not very even, although the survey results are valid and can support further analysis.

1. Uneven distribution of age

   In Chap. 5, the age distribution of the three samples is analyzed and it is found that the age distribution is not very even. The age group of 16–35 is slightly overrepresented. This can be explained by several reasons. First, it is easier to communicate with younger people. In general, younger people are more open, compared to older people. Surveys are not common in China, and it is easier to explain the project to young people, who are curious about new things and have the ability to process information rapidly.

   Secondly, the people of the age group 36–55 are in the business period of their life. They are completely occupied with their career, family, and children's education. Unless they have overwhelming interest in the porpoise protection, they do not have much motivation for attending the workshops.

   Thirdly, dialect is also an important factor, hindering the organization of workshops. In Beijing, it is possible to understand local residents who speak Mandarin, and hence, some old people could be recruited from the neighborhood. In Guangzhou and Nanchang, residents speak their own dialects that cannot be understood. For example, people in Guangzhou speak Cantonese. Only young and highly educated people tend to speak Mandarin. As a result, it was easier to invite younger people to the workshops because of the dialect obstacle.

   Fourthly, the age group (>65 years) is slightly underrepresented in the sample, which is not beyond expectation. The people of this age group (>65 years) are not easily accessible because they are seriously influenced by the news reports from media, where cheating cases on old people are often reported and elderly people are asked not to speak with strangers. For example, some fraudsters fool old people into buying some healthcare products. Communicating with older people was indeed much more difficult. When trying to talk with an old person, he/she refused in a friendly way or went away directly without a word. Consequently, the recruitment of elderly people was difficult.

   Finally, cultural factor also influenced the recruitment. The writings of Confucius served as the foundation of Chinese education for some 2,000 years. Taking Confucianism as the core traditional culture of China, the Chinese always maintain the order of the different hierarchy, senior and junior. If an interviewer is young, he/she does not have enough authority to invite old people, which also influenced the recruitment process. Similarly, if an organizer is older, it is more feasible to invite elder people to attend the workshops.

   Recommendation: hence, for future survey research in China, it is suggested to employ a few older interviewers to get older people to attend the workshops. Also, local translators are greatly recommended to be employed who can help with the recruitment and moderation in the valuation workshops, because local moderators receive more trust from local residents. In addition, it is also suggested that in-person interviews should be employed in the future from the perspective of involving more old people in a survey. To visit an elderly person makes more sense than to invite him/her to come to you in Chinese traditional culture, especially when you are asking for his/her help.

2. Underrepresented less-educated people
   Concerning the distribution of individuals' highest educational level, most respondents hold a bachelor or a master degree. The less educated people, who hold high school degrees or below, are underrepresented. The uneven distribution of the highest educational levels can be explained for a number of reasons:

   First, it is easier to explain the project to highly educated people. The higher the people are educated, the more they are informed about the environment and wildlife issues. Based on the knowledge they already possess, they can understand the porpoise conservation project relatively easily.

   Secondly, the higher the people are educated, the more they care about environmental and wildlife issues. Accordingly, highly educated people are more motivated to attend such workshops on endangered rare species, compared to the less educated people. People's environmental awareness played a great role in determining whether to attend the valuation workshops.

   Thirdly, Beijing and Guangzhou are very large cities and the recruitment took place in the city center. It is less possible to recruit less educated individuals in the city center. Usually, less educated people do physical work in factories located in suburban areas or on farmers in rural areas.

   Recommendation: In order to improve the representativeness of samples, it would be necessary to recruit more less educated individuals also in suburban areas. In addition, when more elderly people participate in a survey, the proportion of less educated individuals can be increased correspondingly. The majority of less educated people in China are 50 years old or older.

3. Family Size
   With regard to the distribution of family size, small families of one or two persons are slightly underrepresented compared with larger families. In particular, only a few respondents answered that their household size is one.

   Respondents may have defined family size in different ways: Firstly, family can be defined by the blood relationship, which is very traditional. Secondly, family can be based on China's household registration system, also called the "*Hukou* system," which is unique in China. The third kind of family definition can be based on cohabitating relationships, which is a new form in recent years. These three kinds of concepts cannot be unified in many people's situations. In other words, it is possible that several family members have a blood relationship and live together, belonging to different household registration systems. It is also common that family members live in different cities when they belong to the same household registration system. So it is very unlikely that people ticked "one-person household."

   In addition, Chinese people have the tradition of respecting the old and loving the young. Though many young couples do not live with their parents, they maintain close contact with them. Grown-up children have the duty to support and help their parents. Some people are used to counting their parents as their family members for their whole life. Furthermore, the Chinese pursue a very strong feeling of belonging. Very few people think that his/her family consists of himself/herself, even if he/she lives alone and has an own budget. As a result, most respondents' answered that household sizes were bigger than two. In this study, family size is verified as an important factor in influencing WTP bids in the subsample of Nanchang.

Finally, the limited sample size cannot guarantee the representativeness of people at all kinds of age and educational levels. Compared to the huge populations in the three cities, the sample size is so small that it is extremely difficult to ensure the equal distribution of socioeconomic characteristics, especially when the recruitment was random and anonymous. It is worth noticing that the limited sample size is a common disadvantage of the group-based approaches, such as valuation workshop. At the same time, the difficulty of having representativeness in a small sample is also prevalent. The unequal distribution of age and educational level is not completely incidental.

Recommendation: It is suggested that the criterion of family size should be clarified in future survey. Excellent strategy of recruitment is essential for optimizing the distribution of sociocharacteristics of respondents to guarantee the representativeness of samples. As a consequence, the survey results can, in turn, be improved.

6.1.2 Protest Responses

It should be noticed that protest responses of both WTP and Non-WTP for the porpoise conservation project account for most among the invalid responses (13.75% in Beijing, 15% in Guangzhou, and 15% in Nanchang). In CVM studies, some respondents protested by simply refusing to pay any amount of money or by stating untrue bids. Protest responses are among the main challenges in a CVM study, because they do not produce genuine WTP. Protest responses in this study may originate from several reasons:

1. Economic development is still given the highest priority in Chinese society. How to deal with the conflict between economic development and the porpoise protection was a hot topic throughout all group discussions. Participants insisted that the local economic development should be central, though they acknowledged that it is important to protect the endangered rare species. In the hypothetical market, a potential Poyang Lake project was discussed. This project is a large-sized hydrocomplex, working on a comprehensive development and utilization of water resources in the lake region. The project receives a lot of attention and support from the Chinese government and local residents. When the conflicts between the water project and the protection of YFPs were put forward during the group discussions, some respondents hesitated to contribute to porpoise conservation, as they were afraid that the Poyang Lake project would be limited for the sake of porpoise conservation.

2. Serious corruption is also one important reason for producing protest responses. China is perceived to be rather corrupt by its public, domestic private entrepreneurs, and foreign investors (Pei 2007). As a result, there is widespread anger with corruption activities in the whole society. People worry about corruption, which can be perceived from participants' statements in the workshops. They were worried that their payment would not be used for the protection project. The bad impact of corruption can be an important cause for people giving protest responses.
3. Failing to save many endangered rare species makes some citizens lose belief in the success of such protection projects. Nearly 40% of mammal species are threatened according to the 2004 *China Species Red List* (McBeath and Leng 2006). Many nature reserves are not operating well for several reasons: firstly, the legislation system for reserve management is not sound. By far, there is no specific legislation on reserves on a legal level issued by the People's Congress. The existing legislation is the *Regulation of Nature Reserves* issued by the State Council. However, this regulation is short of detailed rules, standards, and guidelines (Xue 2000). Secondly, beneficial relationships between nature reserve and local community are not properly dealt with. For example, local farmers have not gained any improvement since reserves were established. Sometimes the local people's lives become even worse than before because their resource use is prohibited after reserves were established. Consequently, poaching and illegal resource harvesting are common in some reserves (Xue 2000). Thirdly, the existing legislation has not been well-performed. The legislation stipulates that no resource development activities are allowed in core zones and buffering zones, but actually many development activities occur in buffering zones and even in the core zones due to many factors such as financial pressure and the administrative system. The competent governmental departments have taken a passive attitude on resource development within reserves. They have neither opposed it nor encouraged it (Xue 2000). In summary, many examples of failure in conserving endangered rare species and in developing nature reserves make residents question the success of this protection project.
4. Some individuals contend that nature reserves are the government's and NGOs' business. Historically, states are responsible for the conservation of endangered rare species, from stipulating relevant laws and regulations to giving funds for establishing, managing, and operating a nature reserve. Since the late 1990s, some NGOs have also started assisting the conservation of rare species. Residents are used to thinking that the species conservation is the government's and NGOs' business and they do not have to contribute to it. It is, therefore, difficult for some residents to accept a proposal of contributing to porpoise conservation.
5. Finally, money is always a sensitive issue in China. People become very cautious when a stranger talks with them about money issues. As a result, some respondents tried to avoid the WTP question. Two individuals refused to answer the WTP question because the question was about a money issue.

Recommendation: Understanding the reasons for protest responses on porpoise conservation, it can be concluded that most reasons are external factors, which are related to the government's behavior and the social situation. Investigators cannot influence these factors, like corruption. The only suggestion is that the moderator or investigator should master the balance between making respondents believe the hypothetical market is realistic and eradicate their worries about being cheated, which can help reduce protest responses to some extent. On the one hand, the moderator should explain the hypothetical market in a way that makes people believe it is realistic so that they could show their true statement. In Chinese thinking, nothing is settled until everything is settled. Some respondents do not take the question seriously if they realize that it is only hypothetical and the project will not really be carried out. On the other hand, it should be emphasized that their WTP is only confirmed and they do not need give their money away. Some respondents were afraid that they are tricked out of money and then refused to answer the question or give an understatement. This point is quite vital for CVM surveys in China.

6.1.3 *Embedding Effect: Finless Porpoises Versus Biodiversity*

Embedding effects (3.75% in Beijing, 3.75% in Guangzhou, and 2.5% in Nanchang) are also detected from some individuals' debriefing answers explaining their WTP amount for the porpoise conservation project. The embedding effect is said to arise when WTP responses do not vary between questions seeking the value of protecting a small element of the environment and a much larger component of the environment. In the survey, some respondents mixed up the conservation of YFPs with the protection of aquatic animals, wildlife protection, ecological protection, and environmental protection. Though respondents were told clearly to aim at porpoise conservation in Poyang Lake, some of them still stated their preference on general wildlife and environmental protection and did not confine it to the conservation of YFPs. This can be explained in several ways, focusing in particular on the Chinese cultural background:

Firstly, it is a kind of typical Chinese logic and thinking. As Chinese words are pictures rather than sequences of letters, Chinese thinking tends toward a more holistic processing of information and emphasizes the big picture over details (Graham and Lam 2003). The key in the Chinese pattern of thinking is linkage. Links are always sought to show the wholeness of life, even when that is not the wholeness of life, even when the whole embraces contrast. Events are linked to other large events that occurred in past times. The unity of human experience with the whole of life is the fundamental philosophical basis for the thinking pattern (Varner and Beamer 2005). Therefore, when individuals were asked about porpoise conservation, they linked it with the more general environmental issues, such as aquatic animals, wildlife, or the ecology of Poyang Lake.

Secondly, there are more news reports and information on environmental and wildlife protection transmitted in media, like TV, the Internet, newspapers, and magazines, compared to the information on the YFPs. In addition, most news reports on one specific endangered species or environmental disaster in China, usually come to end with a general calling on people's attention to nature or ecology. For example,

"*Save our world! Protect our Earth!*" Consequently, residents are more familiar with the general environmental protection, and the protection of all endangered species is more appealing than the conservation of one particular species. Sometimes people think ambiguously that the environmental protection or wildlife protection and the protection of any specific endangered species, like the YFPs, are one thing. In their opinion, as long as the environmental quality of Poyang Lake is improved, the YFPs can live completely safe. Or else, in some people's opinion, to protect YFP is to protect the environment. To maintain the ecology of Poyang Lake is to conserve the YFPs.

Thirdly, the general environmental protection is more closely related with citizens' understanding than porpoise conservation. For example, benefits generated from the improvement of water quality in Poyang Lake are more practical for local residents than those from the increase of the porpoise population. With better water quality in Poyang Lake, fishery can be increased. People feel that the benefits produced from porpoise conservation are a bit far from them.

Finally, more responses of embedding effects are detected in Beijing and Guangzhou than in Nanchang. The farther away people live from Poyang Lake, the less familiar they are with the YFPs. It is verified that participants in Guangzhou or Beijing stated more responses biased by embedding effects because they were less familiar with the YFPs. They might have used the focus on environment in general as a cognitive heuristic. It is possible that distance may increase embedding or familiarity might decrease embedding, which needs confirmation by further research in the future.

However, the occurrence of embedding effects is not always a pervasive feature of CVM studies, if the regional context is clearly communicated to the respondent (Loomis et al. 1993). Clarification of the valued public goods is one decisive step for reducing the occurrence of embedding effects, or at least the magnitude of its effect on WTP. The valued goods should be very clearly defined and their provisions should be made plausible (Carson and Mitchell 1995). Some suggestions are put forward to strengthen the clarification of the valued object, such as using identical scale maps to represent different levels of public goods, applying a different survey method such as an in-home interview (Loomis et al. 1993), careful attention to potential amenity misspecifications, the use of plausible market scenario, the choice of nested goods whose differences are meaningful to respondents, and a sample large enough to reject reasonable differences (Carson and Mitchell 1995).

In this study, the suggestion to reduce embedding effects is to keep emphasizing that the object to be valued is the conservation of YFPs. Based on Chinese holistic thinking, it is also necessary to take considerable efforts to clarify the relationship between porpoise conservation and the improvement of lake ecology in order to reduce or even prevent embedding effects. Respondents should be made very clear about the valuation focus in follow-up research. In this questionnaire, individuals were only explained contents and possible achievements of the conservation project. Additionally, it should also be made clear what cannot be gained from the project. For example, the project cannot help improve the ecology quality of the whole lake. Other wildlife will not benefit from the project directly, except the YFPs.

In addition, respondents biased with embedding effects are not included in the valid-results analysis in this study. It is worth studying in-depth research regarding whether responses biased with embedding effects should be treated as valid and also included in the analysis. For instance, the WTP for different protection levels, such as the ecology of Poyang Lake and wildlife protection, should be tested to differ significantly or not in future research. After all, there is no agreement on the principles that could guide the choice of the appropriate embedding level for the good to be valued (Kahneman and Knetsch 1992). The designer of a CVM survey may be able to determine the estimated value of any good by the choice of a level of embedding in future research.

6.2 Distance Effect

An important aim of this study is to test the distance effect on the WTP. In Chap. 3, the results of the existing research on distance effects are summarized. They indicate that distance decay occurs on some goods, such as water quality, national parks, and wetlands, but not on other goods, such as salmon protection and forest easement programs.[1] In this study, workshops were organized to survey individuals' WTP for the proposed conservation project on the YFPs in Beijing (about 2,000 km away from Poyang Lake), Guangzhou (1,000 km away from Poyang Lake), and Nanchang (located in Poyang Lake region, 50 km away). The results of the study indicate that the mean WTP in Beijing is always the largest among the three subsamples. On the whole, the mean WTP in Nanchang is higher than that in Guangzhou. Similarly, the mean WTP of high income groups (1,500–4,499 RMB per month) is bigger in Nanchang than in Guangzhou. However, the WTP of low income groups (less than 1,500 RMB per month) is bigger in Guangzhou than Nanchang. At first, the mean WTP in Guangzhou is smaller than that in Nanchang, which is consistent with the previous findings, where the WTP decreases with increasing distance. However, the mean WTP in Beijing is biggest among the three subsamples, which contradicts expectations derived from distance decay. It can be concluded that distance effects do not work on the WTP for porpoise conservation in China.

The findings can be explained in several ways: Firstly, regression results show income exerts greatest influence on the WTP, while distance is positively and highly related with income, which is up to the actual economic situation in the three cities. Beijing and Guangzhou are among the most developed cities in China and both hold important economic positions. Residents in Beijing hold the highest income

[1] A forest easement program refers to a new forest management plan, which provides incentives for private forest landowners to manage their forests to meet certain goals, such as sustained timber harvest, permanent protection of forested areas, improving habitat for migratory forest songbirds of conservation concern and reducing the effects of deer browse on tree regeneration (Racevskis 2005).

level, higher than those in Guangzhou. Residents in Nanchang are relatively poor, compared with those in Beijing and Guangzhou. Correspondingly, the average income level of individuals in the Beijing subsample is the highest while that in the Nanchang subsample is the lowest. A high income level decides the high WTP bids in Beijing and the low WTP bids in Nanchang.

Secondly, residents in Beijing are indeed more informed about the environment and nature, compared with people in Guangzhou and Nanchang. As the capital of China, the government is making more efforts on environmental protection and improving residents' environmental awareness. In general, a new environmental education program has been launched from Beijing. At the same time, there are more chances and activities to promote people's environmental awareness, such as the 2008 Beijing Olympics. Because of the Beijing 2008 Olympic Games, a large number of ecological improvements were made, and many steps toward continued environmental protection were taken, such as the increase of green coverage, propaganda of environmental knowledge among citizens. Consequently, individuals in Beijing have a higher environmental awareness and, therefore would like to contribute more to the protection project, which also explains the high WTP in Beijing.

Thirdly, it was found in Sect. 5.5.4.4 that variables *ATTENTION* and *Envi.Group* explain the higher WTP in Nanchang than in Guangzhou significantly. Comparing the subsamples of Guangzhou and Nanchang, there are far fewer participants with environmental group membership in Guangzhou than in Nanchang. Similarly, people in Nanchang care more about the environment and nature than people in Guangzhou, in general. In Guangzhou, many people have the habit of consuming wild animals and like eating animals, like snakes, rats, rabbits, turtles, dogs, cats, and so on, which they regard as natural, tasty, and beneficial to their health. The Southern Weekend newspaper recently cited a survey made by the Guangdong Forestry Bureau that showed that more than half of the people in Guangzhou had eaten wild animals. Reasons cited for the practice showed 45.4 % of the surveyed said they believed wildlife is "nutritious," 37 % were "curious," and 12 % said they did it to flaunt their wealth (*Global Times* 2010). The fact can strongly tell that individuals in Guangzhou care far less about the protection of animals than those in Beijing and Nanchang. Individuals' low environmental awareness definitely explains their low WTP bids.

Fourthly, among the low income groups (less than 1,500 RMB per month), the mean WTP in Guangzhou is higher than that in Nanchang, which is inconsistent with the results of both total valid subsamples and high income groups (1,500–4,499 RMB per month). It is presumed that two reasons are responsible for the result. One reason is the educational level that is found to be highly related with individuals' WTP in low income groups. Many low income individuals in Nanchang are blue collars, working in factories and their highest educational level is only senior high school or junior high school, while all low income individuals in Guangzhou are students and hold the highest educational level of bachelor or master. Another reason is budget constraint. Most low income individuals are students who would have been willing to pay more if they had more money. It is supposed that poor individuals' WTP were distorted by their ATP. Some individuals did state that they would like to pay more when they have a higher income.

Fifthly, since the YFPs are a very rare species almost without any use value, the WTP for porpoise conservation is mostly driven by non-use values. In the survey, respondents also indicated that they care more about the non-use values than the use values of the porpoise conservation (see Sect. 5.2.4). Most of the previous studies, which prove the existence of distance effects, focus on the distance effects on use values. Regarding distance effects on non-use values, Sutherland and Walsh (1985) conclude that distance decay plays a negative but insignificant role on the WTP for option, existence, and bequest value of water quality. That is, WTP for the non-use values of water quality decrease with increasing distance. But the role of distance is not important for determining WTP bids. Concu (2005) found that the non-use values of bushland in Kings Park were distance-independent. Therefore, results stating that the mean WTP of the Guangzhou subsamples is smaller than that of the Nanchang subsample do not contradict the findings of the previous studies. This finding suggests that the values of environmental goods remain stable over distance when the good under investigation is composed of non-use values.

Sixthly, it is possible that the market extent for the YFPs could be quite large, much wider than the radius of 2,000 km, the distance from Beijing to the Poyang Lake region. Among the previous studies, the market extents of some goods were quite large. For example, the market extent of the salmon program is the whole USA, beyond 3,000 miles (equaling 4,800 km) (Loomis 1996). If it is intended to detect the distance effects on the WTP for porpoise conservation, a survey in a wider range may be required, because the WTP for porpoise conservation is mostly non-use value driven, and non-use values exist everywhere.

Finally, it should be noticed that distance effects do not work on WTP for all kinds of environmental goods or services. It is proved that distance-decay effects did not exist on WTP for both use-values and non-use values of all Thames region rivers (Hanley et al. 2003) and for the Salmon program mostly driven by use-values (Pate and Loomis 1997). In addition, some studies come to the same results as this study that distance is positively related to WTP for wetland improvement programs consisting mostly of non-use values and contamination control owing to substitution effects[2] (Pate and Loomis 1997) and for forest easement program due to respondents' biocentric attitudes[3] (Racevskis 2005). Possibly, the WTP for porpoise conservation is distance-independent.

In summary, the results of this study show that distance is not a factor influencing individuals' WTP in China, which implies that the market extent of the YFPs is very large—nationwide or even worldwide.

[2] The substitute variable for the wetlands and contamination control programs was an estimate for acreage of wetlands in California, Oregon, Nevada and Washington (Pate and Loomis 1997).

[3] Biocentric attitude refers to an attitude of respect for nature. People, who hold biocentric attitude, tend to agree that it is important to protect the habitat for forest songbirds of conservation concern even if it means economic losses to forest-based industries (Racevskis 2005).

Besides comparing the WTP in the three cities of China, the mean WTP was also compared between China and Germany. In the subsequent section, possible reasons incurring the great difference of mean WTP in the two countries will be explained.

6.3 Different WTP in Germany and China

An international comparison of the WTP for porpoise conservation was conducted between China and Germany so as to gain an impression of how different Chinese and German preferences for porpoise conservation are. The results indicate that the mean WTP in Germany is much larger than that in China.

Since income significantly influences both WTP in China and Germany, the difference of mean WTP can be explained by income differences. Respondents in Germany earn much more than those in China. The average individual income of respondents in Germany was 500–1,000 Euro per month while that of respondents in China was 2,500–3,000 RMB (about 250–300 Euro) per month. Respondents in Germany, therefore have a greater ability to pay. It can be concluded that income was the most important factor determining respondents' WTP.

Based on the factor income, an important issue of the influence of budget constraint on the WTP comes to light. Consumer theory suggests that the individual's budget constraint strongly influences the demand for a good, and therefore the consumer's WTP (Loomis et al. 1994). Respondents may state WTP amounts that are much higher than the amounts they can actually pay, if they do not consider their income constraints properly (Russell et al. 1995; Ahlheim 1998). The YFPs live in the Yangtze River of China Chinese are more familiar with them and care more about them. Hence, one could assume that the Chinese would state a higher WTP if they were not constrained by their income. However, people were explicitly asked to consider their income constraints before answering the WTP questions for porpoise conservation in the questionnaire, which is consistent with the criterion proposed by NOAA Blue Ribbon Panel on CVM (Arrow et al. 1993). This leads to the survey results that indicate the opposite, namely that mean WTP in China is smaller than that in Germany. Due to the income effects, the WTP might be lower than their true preference. Therefore, the WTP for porpoise conservation might be underestimated due to budget constraints among Chinese respondents.

In addition to income as an important factor, there are some other potential reasons for a different WTP in China and Germany. A comparison of WTP between equivalent income groups in Germany and China still provides evidence that the respondents in Germany offered a higher WTP than respondents in Guangzhou and Nanchang of China, but a lower one than those in Beijing, though the mean WTP bids do not differ statistically. It could be perceived during the workshop discussions that factors such as environmental attitudes, environmental behavior, and environmental education also affect individuals' WTP for porpoise conservation.

1. *Environmental attitude*
   Attitudes do influence behavior (Luzar and Cosse 1998). The NOAA panel specifically recommends using "attitudes toward the environment" to interpret responses to valuation questions (Arrow et al. 1993). Environmental attitudes have been found to be significant factors in behavioral analyses of environmental participation (Lynne et al. 1988), consumer response (Allen et al. 1992; Wilson et al. 1990), and environmental management (Heberlein 1989). Environmental attitudes can be assumed to exist and affect environmental behavior (Luzar and Cosse 1998).

   The Chinese are profoundly influenced by Taoism, Confucianism, and Buddhism. The Taoists suggest that nature can serve as a template, a model, a paradigm, an anchor, a beacon and help us to get back in touch with our natural selves. Some Confucian ideas tell that heaven is human's father and the Earth is human's mother. Human is the filial son and daughter of the universe, which connotes a profound feeling, an all-pervasive care for the world. According to Buddhism, the earth is the mother of all the beings that live upon her. Since all of those beings have the same mother as we do, they are all our own brothers and sisters (Epstein 1999). Due to the traditional influence, the Chinese in general express a positive ecological affect and a high green purchase intention (Chan and Lau 2000).

   By contrast, Germans live largely in the context of Christian belief that man is not simply a part of nature, but God has planned the universe for man's benefit. The universe is not a random collection of atoms strung together by chance, but the creation of God, who is described as everything that is good (Gen 1:31). The fundamental difference of environmental attitudes between the Chinese and Germans can be proved by the following facts: a small proportion of German respondents (31 %) (strongly) agree that the impact that our society has on the environment is so severe that there is very little individuals can do about it. On the contrary, 47 % of Chinese respondents (strongly) agree with this view (National Geographic and Globescan 2009). That is, Germans are more optimistic about their capability of reversing their impact on environment, compared with Chinese.

   In addition, people in Germany have a higher sensitivity to environmental problems. According to a survey conducted by National Geographic and GlobeScan in 2009, the results suggest that 22 % of German respondents are very concerned with the loss of species and habitats, while 20 % of Chinese respondents are very concerned with it (National Geographic and Globescan 2009). In a sentence, Germans and the Chinese are both positive on the environment, but Germans have a slightly stronger sensibility to environmental issues than the Chinese, which can explain to some extent the higher WTP in Germany compared to China.
2. *Environmental behavior*
   Environmental attitude can instruct or even decide environmental behavior. However, there is some discrepancy between environmental attitude and environmental behavior (Ajzen et al. 2004; Sheeran 2002; Meyerhoff 2006): for instance, in China, though some fairly comprehensive environmental laws are in place, their

actual effectiveness on the implementation level is still ambiguous, because people often only concentrate on the benefits derived from rapid economic development, but ignore the urgent threat emerging from ecological deterioration (Chan and Wong 1994). Findings of a survey in Beijing and Guangzhou by Chan and Lau (2000) suggest that the Chinese people's level of knowledge and actual involvement in green purchase is still rather low. Individuals' scores on ecological knowledge is only 0.26 out of 1, while their actual commitment to green purchase is only 2.04 out of 5 (Chan and Lau 2000).

By contrast, individuals in Germany are more active in commitments to the environment. According to a report by the Federal Environment Agency of Germany, 45 % of residents engage voluntarily/honorarily, through voluntary activities in environmental groups, or through participation in payment for particular environmental protection activities and projects. Seventy-six percent of the people, who announce that they are actively engaged in environmental and nature protection, are members of environmental and/or nature organizations. Thirty-three percent of the Germans have contributed some money to environmental protection once, while 39 % of them have paid more than once for it (Kuckartz et al. 2007). Aid to disadvantaged groups, life in harmony with nature, political engagement, and development with fantasy and creativity promote the public commitment in the environment and nature protection to the greatest extent in Germany.

Compared to the small differences of environmental attitudes between Germans and the Chinese, Germans are much more active in environmental commitment. This can explain the higher WTP in Germany. In addition to environmental attitude, environmental knowledge is also an important factor for changing human actions toward the environment. Furthermore, there is a relationship between environmental knowledge and environmental attitudes (Arcury 1990). Environmental knowledge attributes to environmental education that people receive. It is quite difficult to measure and compare respondents' environmental knowledge directly. But different environmental education situations in Germany and China can help explain the gap in environmental knowledge between the Germans and the Chinese.

3. *Environmental Education*

People's ecological attitude and environmental behavior are closely related with the environmental education they receive. In China, formal environmental education began at the college/university level in 1970s and then spread into the K–12 school system (including primary school and high school, 12 years in total) (McBeath and McBeath 2009). Except for the formal education with curriculum, some extracurricular activities are carried out. At the same time, environmental knowledge is also diffused by some environmental NGOs, like Friends of Nature, in the form of seminars, training sessions, public events, such as "Global Day," "World Water Day," and "Green Hope" (Hong et al. 2006). Public participation in environmental protection is also encouraged in many large cities, particularly in Beijing. For instance, there is some propaganda posted in communities to tell people how to save water, how to reduce electricity consumption, and how to separate rubbish. Furthermore, some environmental education programs are transmitted through publishing brochures and mass media. Though much work has been done on the environmental education in China, it is still in a developmental phase. It can be verified by the evidence that the environmental curriculum is still incomplete in the educational system of China, which makes it difficult to environmentally educate students and the public.

At the same time, environmental education in Germany is much more developed. Environmental education takes place in Germany in all sectors of the education system. The most important institutions are, on the one hand, schools, universities, colleges, and research institutes, and on the other hand, adult education centers, the church, trade unions, environmental centers and associations, as well as the mass media (Eulefeld 1995). For example, environmental education is compulsory for all pupils in primary schools within the content of existing subjects. According to the curriculum documents, environmental education should start with the everyday experience of students in their local environment (e.g., school, family, home). On walking tours, teaching walks, or other opportunities students can experience the environment that has emotional importance for them (Stokes et al. 2001). Environmental education makes Germans quite aware of the importance of conserving the environment and wildlife and then commit to environmental protection more actively.

In summary, large income differences play a key role in deciding the different WTP in Germany and China. Besides income, people's environmental attitude, ecological behavior, and environmental education are also important determinants of people's WTP. Obviously, German's sensibility to environmental protection is stronger than the Chinese on the whole. Germans are more active in participating in environment-friendly activities than Chinese. Consequently, Germans' mean WTP is higher than the Chinese' mean WTP.

6.4 Summary

The chapter discussed the factors that might be responsible for the difficulty with the performance of the CVM study on the YFPs in China, in terms of uneven distribution of individuals' socioeconomic characteristics, protest responses, and embedding effects. A high time demand, people's fast-speed life style, and their unfamiliarity with the YFPs and rare ecological knowledge increased the difficulty of recruitment. It was found that large income differences, educational degree, and environmental membership explain the WTP in the three cities of China at a significant level. The high correlation between income and distance weakens the influence of distance on WTP. Similarly, it was confirmed that income plays a major role in determining the different WTP between China and Germany. Besides, there are some potential causes responsible for the different WTP in China and Germany, such as environmental attitudes, environmental behavior, and environmental education.

References

Ahlheim, M. (1998). Contingent valuation and the budget constraint. *Ecological Economics, 27*, 205–211.

Ajzen, I., Brown, T. C., & Carvajal, F. (2004). Explaining the discrepancy between intentions and actions: The case of hypothetical bias in contingent valuation. *Personality and Social Psychology Bulletin, 30*, 1108–1121.

Allen, C., Machleit, K., & Kleine, S. (1992). A comparison of attitudes and emotions as predictors of behavior at diverse levels of behavioral experience. *Journal of Consumer Research, 18*, 493–504.

Arcury, T. A. (1990). Environmental attitudes and environmental knowledge. *Human Organization, 49*, 300–304.

Arrow, K., Solow, R., Portney, P. R., Leamer, E. E., Radner, R., & Schuman, H. (1993). Report of the NOAA panel on contingent valuation. *Federal Register, 58*(10), 4601–4614.

Carson, R. T., & Mitchell, R. C. (1995). Sequencing and nesting in contingent valuation surveys. *Journal of Environmental Economics and Management, 28*, 155–173.

Chan, R. Y. K., & Lau, L. B. Y. (2000). Antecedents of green purchases. *Journal of Consumer Marketing, 17*, 338–357.

Chan, H. S., & Wong, K. K. K. (1994). Environmental attitudes and concerns of the environmental protection bureaucrats in Guangzhou, People's Republic of China: Implications for environmental policy implementation. *International Journal of Public Administration, 17*, 1523–1554.

Concu, G. B. (2005). *Are non-use values distance-independent? Identifying the market area using a choice modeling experiment* (No.WP6M05). Retrieved December 2005, from http://www.uq.edu.au/rsmg/WP/WPM05_6.pdf

Epstein, R. (1999). Pollution and the environment: Some radically new ancient views, in English and Chinese. *Vajra Bodhi Sea: A Monthly Journal of Orthodox Buddhism, 30*, 36–48.

Eulefeld, G. (1995). Environmental education in the Federal Republic of Germany. *History of European Ideas, 21*, 17–29.

Global Times. (2010). *Animal protect activities in Guangzhou*. Retrieved August 26, 2010, from http://www.lifeofguangzhou.com/node_10/node_37/node_85/2010/08/26/128279357679701.shtml

Graham, J. L., & Lam, N. M. (2003). *Negotiating in China*. Retrieved October 13, 2003, from http://hbswk.hbs.edu/archive/3714.html

Hanley, N., Schläpfer, F., & Spurgeon, J. (2003). Aggregating the benefits of environmental improvements: Distance-decay functions for use and non-use values. *Journal of Environmental Management, 68*, 297–304.

Heberlein, T. A. (1989). Attitudes and environmental management. *Journal of Social Issues, 45*, 37–57.

Hong, J., Guo, X. M., & Marinova, D. (2006). Chapter 35 NGOs and environmental education in China. In S. Wooltorton & D. Marinova (Eds.), *Sharing wisdom for our future: Environmental education in action* (pp. 324–332). Sydney: Australian Association for Environmental Education.

Kahneman, D., & Knetsch, J. L. (1992). Valuing public goods: The purchase of moral satisfaction. *Journal of Environmental Economics and Management, 22*, 57–70.

Kuckartz, U., Rheingans-Heintze, A., & Rädiker, S. (2007). *Informationsverhalten im Umweltschutz und Bereitschaft zu bürgerschaftlichem Engagement*. Retrieved September 2007, from Im Austrag des Umwelt Bundes Amtes: http://www.umweltbewusstsein.de/deutsch/2006/download/informationsverhalten.pdf

Loomis, J. (1996). How large is the extent of the market for public goods: Evidence from a nationwide contingent valuation survey. *Applied Economics, 28*, 779–782.

Loomis, J., Lockwood, M., & DeLacy, T. (1993). Some empirical evidence on embedding effects in contingent valuation of forest protection. *Journal of Environmental Economics and Management, 24*, 45–55.

Loomis, J., Gonzalez-Caban, A., & Gregory, R. (1994). Do reminders of substitutes and budget constraints influence contingent valuation estimates? *Land Economics, 70*, 499–506.

Luzar, E. J., & Cosse, K. J. (1998). Willingness to pay or intention to pay: The attitude-behavior. *Journal of Socio-Economics, 27*, 427–444.

Lynne, G. D., Shonkwiler, J. S., & Rola, L. R. (1988). Attitudes and farmer conservation behavior. *American Journal of Agricultural Economics, 70,* 12–19.

McBeath, G. A., & Leng, T. K. (2006). *Governance of biodiversity conservation in China and Taiwan*. Cheltenham: Edward Elgar.

McBeath, J., & McBeath, J. (2009). *Environmental education in China: A preliminary comparaive assessment.* etrieved rom www.ccny.cuny.edu/aacs/Conference_Paper_2009/Jerry_and_Jenifer_McBeath

Meyerhoff, J. (2006). Stated willingness to pay as hypothetical behavior: Can attitudes tell us more? *Journal of Environmental Planning and Management, 49*, 209–226.

National Geographic and Globescan. (2009). *Greendex 2009: Consumer choice and the environment—A worldwide tracking survey*. Retrieved from http://environment.nationalgeographic.com/environment/greendex/

Pate, J., & Loomis, J. (1997). The effect of distance on willingness to pay values: A case study of wetland and salmon in California. *Ecological Economics, 20*, 199–207.

Pei, M. X. (2007). Corruption threatens China's future. *Policy Brief, 55*, 1–7.

Racevskis, L. A. (2005). *Public preferences and nonmarket values for the management of forest ecosystem services*. Doctoral Dissertation, Michigan State University

Russell, G. L., Miller, J. R., & Rind, D. (1995). A coupled atmosphere-ocean model for transient climate change studies. *Atmosphere-Ocean, 33*, 683–730.

Sheeran, P. (2002). Intention-behavior relations: A conceptual and empirical review. *European Review of Social Psychology, 12*, 1–36.

Stokes, E., Edge, A., & West, A. (2001). *Environmental education in the educational systems of the European Union*. Synthesis Report committed by the Environment Directorate-General of the European Commission. Retrieved from http://citeseerx.ist.psu.edu/viewdoc/download?doi=10.1.1.102.9715&rep=rep1&type=pdf

Sutherland, R. J., & Walsh, R. (1985). Effect of distance on the preservation value of water quality. *Land Economics, 61*, 281–291.

Varner, I., & Beamer, L. (2005). *Intercultural communication in the global workplace* (3rd ed.). Boston: McGraw-Hill Higher Education.

Wilson, T. D., Lisle, D. J., & Kraft, D. (1990). Effects of self-reflection on attitudes and consumer decisions. *Advances in Consumer Research, 17*, 44–47.

Xue, D. Y. (2000). *Resources management within nature reserves in China*. Papers presented on Economics, Ecology and the Environment, Working Paper No. 51, the University of Queensland.

Chapter 7
Conclusions

Abstract The overall aim of this research was to estimate the non-market values of the YFPs in Poyang Lake, by eliciting thoughtful and informed values from a number of people. Three important issues were put forward in accordance with findings from literature on economic theory, CVM, and the YFPs. These were to:

- Find a way that enables individuals to construct preferences for an unfamiliar species, namely the YFPs
- Test distance effects on WTP for porpoise conservation
- Detect budget constraints on WTP

This concluding chapter is organized in two parts: Firstly, the results of the methodological and the policy-related implications of the study are summarized and secondly, recommendations for future research are suggested.

Keywords Valuation workshop • Distance decay • Budget constraints • Policy-related implications • Recommendations

7.1 General Conclusions

Methodological implications that were addressed in their study include the adequacy of a group-based approach—the valuation workshops (Sect. 7.1.1), absence of distance effects on environmental goods that are mainly driven by non-use values (Sect. 7.1.2), and the important role of budget constraints (Sect. 7.1.3).

7.1.1 Adequacy of the Valuation Workshop

Based on an extensive literature review, various survey methods were compared with regards to conventional features, such as representativeness, sample size, time

Y. Dong, *Contingent Valuation of Yangtze Finless Porpoises in Poyang Lake, China*,
DOI 10.1007/978-94-007-2765-6_7,

cost and budget constraint, and features essential for preference construction. After a comprehensive analysis of the pros and cons of different survey methods, the valuation workshop was selected as the survey method. In this study, porpoise conservation is relatively new and complex. Though the disadvantage of the method is its limited sample size, the method was still chosen as the most proper one in this study, compared with other survey methods.

The final survey results prove that the valuation workshop is suitable for valuing porpoise conservation. Evidence can be found in the regression analyses and the fact that very few non-responses to the WTP question are produced. Participants cooperated with the survey actively and most respondents would like to contribute to the porpoise conservation project. Except for a few biased responses, such as protest responses, strategic bias, embedding effects, and warm glow effects, the WTP responses were confirmed to be valid and the valid WTP response rate was high enough to conduct a WTP analysis. Therefore, it can be concluded that the valuation workshop provides a friendly environment and open atmosphere where individuals can construct their preference on both use values and non-use values of environmental goods and services. The valuation method is suitable for valuing the unfamiliar and complex goods, such as rare species.

The application of the valuation workshop was successful. However, there were a lot of difficulties with the recruitment for the following reasons. Firstly, participants need to spend 3 or 4 h both on the way for the group meeting and during the group discussions, which was too demanding for some people. Chinese cities are usually large and traffic jams are very serious, especially in megacities like Beijing and Guangzhou. Secondly, the arranged time cannot be flexible. The valuation workshops usually happen at night and people go there after work. Working overtime is always very prevalent in China. Many people could not be motivated to attend the valuation workshops after 1 day's hard work, especially when the monetary incentive is not high enough. Finally, it is quite difficult for a young moderator to invite old people to attend the valuation workshops due to the traditional Chinese cultural values. Consequently, the samples are not representative on some socioeconomic characteristics, such as age and educational level.

7.1.2 Absence of Distance Decay

Distance decay is a challenging issue in CVM research. Until now lots of studies have investigated distance decay but the conclusions from these studies are not consistent. This study also tests distance effects on WTP for porpoise conservation.

The survey was carried out in the cities of Beijing, Guangzhou, and Nanchang. It was found that distance effects do not work on WTP bids for porpoise conservation in Poyang Lake. This can be supported by the following evidence:

Firstly, the mean WTP of the subsample in Beijing is biggest among the three subsamples, which contradicts the theory of distance decay. Distance is closely and positively related with the WTP bids of the three subsamples. However, the linear relationship between distance and WTP is deemed unreliable, based on the very low R^2 value.

Secondly, the mean WTP in Nanchang is bigger than that in Guangzhou, which is consistent with the theory of distance effects. However, the variable *DISTANCE* is not found to be highly correlated with WTP bids in Guangzhou and Nanchang, which means that distance does not exert an influence on WTP responses in Nanchang and Guangzhou.

Finally, it is found that the mean WTP in Beijing > the mean WTP in Guangzhou > the mean WTP in Nanchang for the low income groups (less than 1,500 RMB per month); while the mean WTP in Beijing > the mean WTP in Nanchang > the mean WTP in Guangzhou for the high income groups (1,500–4,499 RMB per month). Moreover, distance is closely related with the WTP bids of low income groups but not with the WTP bids of high income groups. However, the linear relationship between distance and the WTP bids of low income group is judged to be unreliable due to the low R^2 value. Until now, it can be concluded that distance does not influence the WTP for porpoise conservation.

The absence of distance effects on the conservation of YFPs has some methodological implications. The WTP for environmental goods that is mostly driven by non-use values does not seem to be influenced by distance. That is, distance does not affect individuals' preference on porpoise conservation, because non-use values can be enjoyed everywhere equally. Consequently, the market extent of porpoise conservation can be very large. It can likewise be nationwide and can even be extended to be worldwide. The benefits from the YFP protection can be consumed by people living far away from Poyang Lake. Therefore, the sampling frame for valuing porpoise conservation can be the entire population of China, the sample is not only confined to the local residents in the Poyang Lake region. Finally, the aggregation of TEV of the YFPs does not need to consider distance. That is, WTP bids in different places should be entitled with the equal weights when calculating the aggregation. All the Chinese population's WTP should be included in the aggregation. Likewise, future CV studies on endangered rare species in China do not need to test distance effects anymore and can take samples anywhere for the researchers' convenience.

7.1.3 Important Role of Budget Constraints

So far, there are very few international comparisons of CVM studies. The mean WTP for porpoise conservation was compared between China and Germany. The YFPs are the only unique porpoise species endemically adapted to the Yangtze River in China. Therefore, they are not only important for the Chinese, but also for other people all over the world. Germany was taken as an example in this study. To check Germans' preference for the YFPs is very meaningful for examining their importance within a wider regional range.

The mean WTP is much higher in Leipzig-Halle, Germany than those in the three cities of China. It is also verified that the mean WTP in Germany differs from that in China, statistically. Furthermore, the mean WTP bids are compared among the subsamples of the same income level (less than 500 Euros per month). The results show that the mean WTP in Beijing > the mean WTP in Leipzig-Halle > the mean WTP in Nanchang > the mean WTP in Guangzhou. These four mean WTP bids of income groups of less than 500 Euros do not differ from each other significantly.

It is confirmed that income effects play a decisive role in determining WTP bids in both China and Germany. Individuals in Germany hold higher income levels than those in China. Consequently, German respondents are able and willing to pay significantly higher amounts of money than Chinese respondents on the whole. However, German respondents did not state significantly higher WTP than the respondents of the same income group (less than 500 Euros per month) in Guangzhou and Nanchang, which proves the income effects again.

YFPs live in the Yangtze River of China and it was expected theoretically that the Chinese would care more about them than Germans and hence would state higher WTP than Germans. As a matter of fact, Germans state a higher WTP than the Chinese. One possibility for the unexpected results is that the Chinese respondents' WTP are determined by their income constraints. That is to say, the Chinese may care more about the YFPs than Germans in fact. However, budget constraints led to the reduction of their stated WTP to some extent. As a consequence, their stated WTP might be lower than their actual WTP.

This can be a common problem with economic valuation of environmental goods in developing countries, where people have low incomes in general. When they are asked to state their preferences on environmental goods by answering WTP questions, they stated WTP bids are possibly distorted by their budget constraints and are, therefore, much lower than their actual WTP bids; though they do care about the environmental and nature issues a lot. This problem should be paid attention to in future CV studies in developing countries.

7.1.4 Policy-Related Implications

This study can provide citizens with the possibility to express their preference on environmental good and legitimize the public opinion on public resource allocation, which is very important for realizing the optimal allocation of scare public goods, such as rare species. In this study, respondents expressed their own attitudes and preferences for the porpoise conservation, and even the whole environmental protection. In the trade-off between economic growth and ecological use, economic growth is often predominant. From now on, governments and decision-makers should take individuals' preference for healthy life into account when they make decisions regarding economic development.

Based on the current high mean WTP bids, it can be predicted that the benefits of the porpoise conservation should be appreciably high. The results of economic valuation can make the non-governmental organizations, who are interested in protecting YFPs, and the government, who plays a key role in protecting YFPs, well-informed about their values. The great values of YFPs should be integrated into the cost-benefit analysis of future development projects in Poyang Lake. It can present an important guidance for political decision on some development projects in Poyang Lake. However, it is an open question that it is an open decision to establish a nature reserve special for YFPs in Poyang Lake.

In addition, the absence of distance effects on the porpoise conservation tells that the benefits from the porpoise conservation can be enjoyed all round China. Therefore, not only local government in the Poyang Lake region but also the state government and the regional government outside the Poyang Lake region should take the responsibilities on the YFP protection. Specifically, governments of different levels and in different places should contribute to and appropriate some funds for the porpoise conservation.

7.2 Recommendations for Future Research

These study results can provide both a practical and theoretical contribution to the understanding of the significance and value of porpoise conservation in Poyang Lake. On the whole, the study shed light on people's WTP for porpoise conservation. However, there are still some disadvantages with it.

Firstly, some difficulties occurred during the application of the CVM, which leads to the uneven distribution of some socioeconomic characteristics and some biased responses. For future CVM research in China, it is recommended that a different survey method should be employed, such as in-person interviews or household drop-offs. Some special elements in the design of the CVM should be taken into account, such as employing some local investigators, conducting the survey in the special communities for the elderly people, and doing the survey both in city center and suburban areas.

Secondly, it is estimated that individuals' WTP in China are biased by their budget constraints, which should be reflected in follow-up research. People can be asked to state their WTP and ability to pay for the same valued object so that the relationship between WTP and the ability to pay can be explored. In this way, it can be judged whether WTP is biased by the ability to pay. Moreover, it is not clear how much an individuals' WTP is biased by his/her budget constraint and it should be examined by conducting some economic experiments in future studies. For example, WTP should be asked in two situations. In one situation, respondents are reminded of their ability to pay. In another situation, respondents are told not to consider their ability to pay. Comparing the WTP bids in the two situations can explain how much WTP is biased by the budget constraints. The test of bias caused by budget constraints should be conducted in all the CVM studies in developing countries, not only in China.

Finally, since it is still an open question whether the establishment of a national nature reserve should be put in practice; this study provided no analyses of different alternatives, which should be conducted in future research.

Appendices

Appendix A Demographic Characteristics of Participants in Focus Groups

Table A.1 Demographic characteristics of participants in Shentangyu Village of Beijing

	Occupation	Age category	Educational level	Gender	Family members	Monthly income (before tax) Unit: RMB
1	Farmer	36–45	Primary school	Female	4	<500
2	Farmer	46–55	Primary school	Female	3	<500
3	Farmer	26–35	Junior high school	Female	More than 4	<500
4	Farmer	36–45	Junior high school	Male	3	500–1,499
5	Farmer	36–45	Junior high school	Female	3	<500
6	Farmer	26–35	Senior high school	Male	3	500–1,499
7	Farmer	56–65	Junior high school	Male	4	<500
8	Farmer	36–45	Junior high school	Female	3	500–1,499
9	Farmer	56–65	Primary school	Female	1	<500
10	Farmer	56–65	Primary school	Male	More than 4	<500
11	Farmer	36–45	Junior high school	Female	2	500–1,499
12	Farmer	46–55	Junior high school	Female	3	<500
13	Farmer	46–55	Junior high school	Female	1	500–1,499
14	Farmer	46–55	Senior high school	Female	2	<500

Note: 1 RMB = 0.109342 EUR

Y. Dong, *Contingent Valuation of Yangtze Finless Porpoises in Poyang Lake, China*,
DOI 10.1007/978-94-007-2765-6, © Springer Science+Business Media Dordrecht 2013

Table A.2 Demographic characteristics of participants in Nanbeigang village in the Poyang Lake region

	Job	Age category	Educational level	Gender	Household members	Monthly income (before tax) Unit: RMB
1	Farmer	36–45	Junior high school	Male	4	<500
2	Farmer and fisherman	56–65	Primary school	Male	Over 4	1,500–2,499
3	Farmer and fisherman	56–65	Illiterate	Male	4	<500
4	Farmer	36–45	Primary school	Female	4	1,500–2,499
5	Farmer	56–65	Primary school	Male	2	<500
6	Farmer	56–65	Junior high school	Male	4	<500
7	Farmer and fisherman	26–35	Primary school	Female	4	500–1,499
8	Farmer	36–45	Primary school	Female	Over 4	1,500–2,499
9	Farmer	46–55	Junior high school	Male	Over 4	<500
10	Farmer	46–55	Junior high school	Male	3	1,500–2,499
11	Farmer	36–45	Primary school	Female	4	1,500–2,499
12	Farmer	56–65	Primary school	Male	Over 4	500–1,499
13	Administrator	26–35	Bachelor degree	Female	Over 4	1,500–2,499
14	Farmer	26–35	Primary school	Female	3	1,500–2,499
15	Member of village committee	26–35	Senior high school	Male	4	500–1,499
16	Farmer	46–55	Primary school	Female	4	500–1,499

Note: 1 RMB = 0.109342 EUR

Appendix B Summary of Focus Group Findings

B.1 People's Knowledge and Perception of Yangtze Finless Porpoises

In Shentangyu village, none of the focus group participants had heard of Yangtze finless porpoises before. People felt that they were very far away from porpoises. Several participants said: "*I have never been to Yangtze River and I have never seen the YFPs so far.*" After they read the information folder, they gained the general impression of YFPs and knew how they look like. It attracted many participants' attention that YFPs are unique in China. "*They only live in Yangtze river. They are very special.*" One person said: "*They are very cute and big.*" Another person said: "The *YFPs are similar with the river dolphins*". Indeed there is more information on river dolphin than the YFPs. "*River dolphins have died out and therefore we should pay more attention to the YFPs. If we do not protect them now, porpoises will also die out.*" Then most of participants concluded that the YFPs were worthy of protection.

In Nanbeigang village of Poyang Lake region, almost all the focus group participants heard of the YFPs, and eight of them even once saw the YFPs in Poyang Lake. Fishermen, who saw porpoises in the Poyang Lake when they fished, stated some information about the YFPs as follows:

> "There are many YFPs in the mouth of Poyang Lake. Porpoises often come out of the water surface."
>
> "When there is north wind, porpoises will come out and face windward. There is a local saying in Hukou county that 'River pigs worship wind'."
>
> "It is easy to see the YFPs in Poyang Lake from July to December every year."
>
> "The YFPs can serve as a weather presumption well. When the YFPs jump out of water, the weather will change badly. Probably it is badly windy."
>
> "I am a fisherman for a long time. I have often seen them but I do not know how many there are on earth. Through the focus group I get more knowledge of porpoises, including the amount, distribution, and population change."
>
> "Grease of the YFPs is a good medicine for curing burned skin."
>
> "There are less and less porpoises in recent years."
>
> "There are lots of the YFPs near the two bridges, namely the railway bridge and the highway bridge, which lie near Hukou county."
>
> "They are very clever. They only come out in bad weather."
>
> "I love porpoises. They can tell me where there are lots of fish, like 'hunting dog'."
>
> "The YFPs are human's good friends."
>
> "I am a farmer. I only know the YFPs but I am not familiar with them. There is too little information on porpoises."

According to their answers, it can be concluded that the participants had good knowledge of behavioral habit and distribution of the YFPs. They did not have much knowledge of porpoises' uniqueness and how their population changed. They assured that they did know more about the YFPs after the focus group. They came to one common conclusion that the YFPs were very important and it was a great honor for local residents to live near them. Especially the fishermen, who had lots of opportunities to see the YFPs, held some personal affection on porpoises.

B.2 People's Views Toward to Protection of Yangtze Finless Porpoises and Reaction to Threats Faced by Them

Participants were provided with the information on the protection status of the YFPs and threats faced by porpoises. In the section on protection status of the YFPs, population change and density change were introduced. Participants were impressed about the great decrease of the YFPs population and the serious threats encountered by the YFPs. In addition, the specialty of the YFPs was explained in the information folder in order to make participants understand why the YFPs need urgent protection.

In *Shengtangyu*, a village of Beijing, people showed some sympathy mixed with innocence with the threats faced by the YFPs.

> "It is a pity that the YFPs are facing serious threats."
>
> "As for the illegal fishing and overfishing, government should tell local fishermen to stop these actions."

With regarding to the protection of YFPs, all the participants confirmed that they did support the protection project on the ground that the YFPs were the Second Category of National Key Protected Wildlife Species in China. Most of them still insisted that government should take the main responsibility for the conservation of YFPs.

> "Though the YFPs are so important that government should upgrade them into the First Order of National Key Protected Wildlife species in China as soon as possible."
>
> "Government should make more efforts to spread the information of the YFPs to more citizens. At least, we did not know it at all before."
>
> "What we can do is very limited. Government should play the key role in protecting the YFPs."
>
> "If the central government wants to carry out a protection project, which requires my contribution, I would like to perform my obligation. Or else, I do nothing, neither threats nor protection."

For sure, some people still would like to contribute to the conservation of Yangtze finless porpoises. Several participants expressed their standpoints.

> "Now economy is developing at a fast speed and our living standard is improving greatly. But environmental pollution is more and more serious and lots of rare species are dying out. Human beings cannot live alone on the Earth. I would like to try my best to protect the YFPs."
>
> "I cannot offer any detailed protection proposals, which should be suggested by experts. But I would like to give some specific supports to the protection project, such as money donation."

In *Nanbeigang*, a village of Poyang Lake region, people mainly concentrated on water pollution because they thought water pollution was a serious threat to both the YFPs and fishery. Local residents complained a lot about water pollution.

> "Lots of waste water is delivered here from factories in Jiujiang city on the sly. They do not ask for our permission."
>
> "There is a new-built crematorium nearby the Poyang Lake. We are afraid the waste water from crematorium would be discharged to the Lake before being treated."
>
> "There are several chemical factories near the lake. Much waste water is discharged into the lake and causes water pollution."

It has to be noticed that fish catch is decreasing due to water pollution and then great economic loss is caused to fishermen. What is more, farmers also suffer a loss in crop production because they use the wasted water for irrigation. That is why people focused on water pollution. When participants were asked about illegal fishing and overfishing, they argued that they never did overfishing or fishing illegally. Since 2002, fishing is banned from April to June. People stated that there exist strict regulations and they would never risk fishing by breaking the law. They said that they do fishing with fishing permits, which were issued by Administration of Fishery in Jiangxi Province. When they were asked about the sand mining, they told that bosses of sand mining companies were all from other places, not local residents. They knew little about sand mining. In their opinion, water pollution was most serious threat to the YFPs.

Concerning the protection of YFPs, participants showed great interest and support. With reference to the specific protection measure, they suggested that waste water should be controlled strictly. If possible, some chemical factories should be removed.

> "Poyang Lake is my hometown and the YFPs are my friends. I hope that porpoises could be protected."
>
> "If chemical factories could be removed away, lake water would be clean. As a result, there will be more and more YFPs."
>
> "We hope the lake water could keep clean forever. Then fish catch keeps big and the YFPs could have enough food."
>
> "If government can work on waste water treatment well, I would like to donate 1,000 RMB per year to the YFPs protection project. Or else, I would do nothing."

B.3 Acceptance of the Hypothetical Market

In the hypothetical market, a nature reserve was proposed to be built up in Poyang Lake. People were asked to elicit their preferences for the YFPs by way of *Willingness-to-Pay* for the establishment of the nature reserve. It contains three parts: (1) the suggested establishment of a nature reserve, (2) the payment vehicle, and (3) the WTP question.

B.3.1 Establishment of a Nature Reserve

From the perspective of protection, the establishment of a nature reserve is an important protection measure, which gained agreement from almost all respondents. No individual was against the protection measure. That is, to build up a nature reserve in Poyang Lake is reasonable and acceptable for residents. Especially in Shengtangzu, focus group participants voted for the protection measure, because Poyang Lake was far away from them and the establishment of nature reserve would produce no negative effect on them. On the contrary, participants thought that some benefits would be brought about to them. Participants took pride in the YFPs, the

Second Category of National Key Protected Wildlife Species, which are almost as important as giant pandas. Participants expressed that they definitely supported the nature reserve project.

> "It's all the Chinese's honor if the YFPs could be well-protected. After all, the YFPs are endemic in Yangtze river in China."

Participants in Nanbeigang, a village of Poyang Lake region, also voted for the protection project of establishing a nature reserve in Poyang Lake.

> "It is our pleasure to establish the nature reserve in Poyang Lake, which indicates that lake water is much cleaner than other parts of Yangtze River system. What's more, it also proves that the ecology here is good."

At the same time, residents in Poyang Lake also expressed their concerns with the establishment of nature reserve in Poyang Lake as follows:

> "If a nature reserve would be built up here, could all the chemical factories be moved away? If so, it should be a great idea."
>
> "If a nature reserve would be built up, would all the illegal sewage discharge be forbidden?"
>
> "If a nature reserve would be built up, fishery would be limited to some extent. Could we get some compensation?"
>
> "Sand mining has serious influence on the protection of the YFPs. What measures would be taken to stop sand mining? How do you deal with the conflicts between sand mining and the YFP conservation? Do you have any good plans?"

B.3.2 Payment Vehicle

In the hypothetical market, income tax was proposed as the way of collecting fund for the protection project on the ground that the money supporting environmental protection normally comes from taxation. Income tax is the cheapest and most reliable way of collecting funds for a long time. When the focus groups participants were asked whether they would like to pay for the protection project in additional taxation, 23 of 30 participants answered "No." Two participants answered "Don't know." Only four participants answered "Yes." One person did not answer the question. At the same time, participants also gave their explanation as follows:

> "Recruitment of income tax would increase our financial burden."
>
> "Agricultural tax was exempted. The money should be offered by enterprises."
>
> "Recruitment of income tax would increase our financial burden. Voluntary donation is better."
>
> "Farmers' burden should be reduced according to policy executed by Rural Tax-Fee Reform Taskforce of the State Council. Tax should be levied on companies and factories."
>
> "I do not have money."
>
> "There are too many chemical factories around the lake, which lead to water pollution."
>
> "I suspect whether the money will be used for the protection project. I do not like corruption."

It has to be noticed that the focus groups were organized in villages and all participants were farmers. There is a new Rural Tax- Free Reform taking place in China, currently. The agricultural tax, China's most ancient tax category, started to be collected in 594 BC. From that time, agricultural tax has existed for 2,600 years in China with dominant rural economy. During the more than 2,000 years, agricultural tax was always the main source of the country's revenue. Since the founding of the People's Republic of China in 1949, agriculture has made great contribution to the country's economic development. In 1953–1985 period, Chinese government purchased grains, cotton, and other agricultural products with unified prices, which were much lower than the prices in the free market, to save money for developing the industry. In this way, Chinese farmers contributed 600–800 billion RMB (about 75–100 billion US dollars) to the country's industrialization. Since the policy of reformation and openness in the 1980s, the gaps between city and countryside, urban citizens and rural residents were widened over recent years. Therefore, solving issues concerning agriculture, countryside, and farmers have turned to be the urgent task for the Chinese government. The ratio of urban to rural income per capita reached 3.31:1 in 2007. It is now time that industry should support the agricultural sector. Therefore, the agricultural tax was exempted since January, 2006, which was of significant meaning to Chinese farmers.

Rural tax-free reform not only releases farmers' financial burden but also lessens farmers' psychological burden in some sense. Correspondingly, to put forward income tax here makes farmers more or less annoyed. As participants recommended, donation or trust fund should be used for the future questionnaire survey. A different way of collecting fund makes them feel more comfortable. It is a kind of psychological factor or phenomenon.

B.3.3 WTP and Its Potential Problems

Among the 30 focus group participants, 16 participants were prepared to pay in principle. Only 11 participants told how much they were willing to pay for the protection project exactly. WTP bids range from 10 to 1,000 RMB. Five participants wrote down "*It depends on my actual income every year. I do not know.*" The remaining 14 participants just left the space blank.

The low response rate can be attributed to several reasons. Firstly, a logic mistake was made in asking WTP questions. The question is:

Q.3 If you answered "Yes," what is the most that your household would be willing to pay in additional taxation each year toward the program to increase the populations of finless porpoises?

Before you answer, please remember that

- The money raised would only go toward protection of YFPs in Poyang Lake
- It would not pay for Yangtze finless poises protection in other areas in China or any other environmental program.
- If the total amount raised in extra taxes was not enough, then the program will have to stop.

The most I would be willing to pay each year for the protection of YFPs in Poyang Lake is:

50 RMB □	100 RMB □	150 RMB □	200 RMB □	300 RMB □
400 RMB □	500 RMB □	600 RMB □	700 RMB □	800 RMB □
900 RMB □	1,000 RMB □	Other __________ RMB/Year		

That is, participants are required to answer WTP question only when they agree with the income tax as the payment vehicle. According to the above statistics, 23 participants disagreed with the income tax. Against this background, it is understandable why so few respondents gave their specific WTP bids.

Secondly, the project is new for the Chinese. They felt strange about the WTP question. They insisted that such protection projects on rare species belong to governments' affair. Rare species are public goods and there is neither market nor price for them. It is indeed difficult for residents to decide WTP.

Thirdly, it was mentioned that the project would last for 50 years and it meant that people need to pay for it every year. Farmers' income changes year by year, which means that participants' ability to pay is changing year by year. They seldom know how much they earn in the following year. Especially, during past 30 years, since the policy of reformation and openness, farmers' living standard has been improved greatly. All at once, people's environmental awareness is also heightening gradually. So, it is possible that participants' maximum willingness to pay varies from year to year in the future. It is explicable that people gave the answer that *WTP depends on my actual income*. Hence, WTP response rate was low.

Finally, it is undeniable that many people do not want to pay for the protection project. People's environmental awareness is still not high enough. They indeed know some rare species should be protected. They think what they can do is not to hurt those species. When they are asked to pay for the protection project, they hesitate. In particular, farmers live on the ground level of the whole society and they are the people who really need help, in their opinion. It is a little too demanding to ask them to pay for porpoises' protection.

In conclusion, the hypothetical market needs modifying. Payment vehicle and elicitation format should also be changed. Moreover, participants should be encouraged to consider about their maximum willingness to pay in a future survey and it should be emphasized that even very low payment is also acceptable in order to wipe off participants' psychological pressure when they could only afford low payment.

Appendix C Summary of Interviews with Experts

Interviewee list:

Professor *Wang Ding*, Institute of Hydrobiology, Chinese Academy of Science, Wuhan, China

Associate Professor *Wang Kexiong*, Institute of Hydrobiology, Chinese Academy of Science, Wuhan, China
Dr. *Zhao Xiujiang*, Institute of Hydrobiology, Chinese Academy of Science, Wuhan China
Secretary-General Professor *Dai Nianhua*, Jiangxi Academy of Science, Ecological Society of Jiangxi Province, Nanchang, China
Huang Xiaoping, Associate Director of Jiangxi Administration of Fishery Bureau, Nanchang, China
Wu Shuyuan, Section Chief of Resources & Environmental Protection, Jiangxi Administration of Fishery Bureau, Nanchang, China.
Professor *Chen Xiaoling*, Jiangxi Normal University, Nanchang, China
Professor *Liu Ying*, Jiangxi Normal University, Nanchang, China

Time and place of interviews:	16–20 March 2009	Wuhan, China
	23–27 March 2009	Nanchang, China

Appendix D Questionnaire

Some general questions about environment and nature (please cross your answer among the following choices!)

Q.1 Programs about the environment and nature are often on TV, internet, and radio. Please indicate, which statement most accurately reflects your own opinion about these programs (Cross one option only)
- I make a special effort to watch/listen **(1)**
- I watch/listen to them when I can **(2)**
- I watch/listen to them if there is nothing better on **(3)**
- I hardly ever watch/listen to them **(4)**

Q.2 As citizens we have priorities to determine the policies that we would most like the government to follow. Please rank your own priorities from 1 (most important) to 5 (least important)
- Providing medical and endowment insurance in rural area ( )
- To lower housing price ( )
- Solving the graduate unemployment issue ( )
- Education ( )
- Environment and Wildlife protection, such as YFPs ( )

Q.3 In what way, do you think, should endangered rare species be protected? *(Please cross only one option)*
- Increase their population **(1)**
- Keep their present population **(2)**
- Keep several in zoo for sightseeing **(3)**
- Leave them alone **(4)**
- I don't care **(5)**
- I don't know **(6)**

(continued)

(continued)

Q.4 It is important to take account of the environment when decisions regarding economic development programs are made	
– Strongly agree	**(1)**
– Agree	**(2)**
– Don't know	**(3)**
– Disagree	**(4)**
– Strongly disagree	**(5)**
Q.5 We should invest in the environment so that the next generation may benefit from the plants and animals on Earth	
– Strongly agree	**(1)**
– Agree	**(2)**
– Don't know	**(3)**
– Disagree	**(4)**
– Strongly disagree	**(5)**
Q.6 Animals such as YFPs have a right to exist, even though they may be of no direct use to mankind	
– Strongly agree	**(1)**
– Agree	**(2)**
– Don't know	**(3)**
– Disagree	**(4)**
– Strongly disagree	**(5)**

Here we would like to ask you a few questions about yourself so that we can understand your answers better. All your answers will be kept confidentially and anonymously.

Q.1 What is your job?	
______________________________	
Q.2 Are you a member of any of the following environmental groups/charities?	
– Friends of Nature	**(1)**
– WWF	**(2)**
– Fiends of the Earth	**(3)**
– Global Village of Beijing	**(4)**
– Wildlife Conservation Society	**(5)**
– Royal Society for Prevention of Cruelty to Animals	**(6)**
– Greenpeace China	**(7)**
– International Fund for Animal Welfare	**(8)**
– WildAid	**(9)**
– International Crane Foundation	**(10)**
– Any other similar groups ---------------	**(11)**
– I do not belong to any such group	**(12)**
Q.3 What is your age category?	
– <16	**(1)**
– 16–25	**(2)**

(continued)

(continued)

– 26–35	**(3)**
– 36–45	**(4)**
– 46–55	**(5)**
– 56–65	**(6)**
– 66–75	**(7)**
– 76–85	**(8)**
– >85	**(9)**
Q.4 What is your highest education level so far?	
– Illiterate	**(1)**
– Primary school	**(2)**
– Junior high school	**(3)**
– Senior high school	**(4)**
– Bachelor	**(5)**
– Master	**(6)**
– Doctor	**(7)**
Q.5 Are you?	
– Male	**(1)**
– Female	**(2)**
Q.6 Which ethnic group do you belong to?	
– Han	**(1)**
– Zhuang	**(2)**
– Manchu	**(3)**
– Hui	**(4)**
– Miao	**(5)**
– Other:____________________	**(6)**
Q.7 How many members are there in your household?	
– 1	**(1)**
– 2	**(2)**
– 3	**(3)**
– 4	**(4)**
– Over 4	**(5)**
Q.8 What is your approximate monthly gross income (before tax)? (*Please remember that your answers are anonymous.*)	
– <500 RMB	**(1)**
– 500–1,499 RMB	**(2)**
– 1,500–2,499 RMB	**(3)**
– 2,500–3,499 RMB	**(4)**
– 3,500–4,499 RMB	**(5)**
– 4,500–5,499 RMB	**(6)**
– 5,500–6,999 RMB	**(7)**
– 7,000–8,999 RMB	**(8)**
– 9,000–9,999 RMB	**(9)**
– >10,000 RMB	**(10)**

D.1 Information Folder on Yangtze Finless Porpoises

D.1.1 Description

- Mammal
- Maximal of 200 cm and 100 kg
- Navigate with sonar system
- Feeds on fish and shrimp
- They are the only porpoises adapted to freshwater
- Live expectancy: approximately 20 years

D.1.2 Distribution

YFPs live in the lower and middle reaches of Yangtze River, where it ranges 1,600 km upstream as far as the Three Gorges Dam above Yichang, including Poyang Lake and Dongting Lake and their tributaries, the Gan river and Xin river (Fig. D.1).

D.1.3 Protection Status of Yangtze Finless Porpoises

Populations Change

As you can see on the graph (Fig. D.2), the population of YFPs is decreasing rapidly.

At present there are ~1,800 individuals living in the Yangtze River system, about 400 individuals living in Poyang Lake and 100–150 porpoises living in Dongting Lake.

It is estimated that they will die out in 25–100 years.

Density Situation

0.02 porpoise/km now in the section from Yichang to Ezhou (Fig. D.3). Scientists predict the density to decrease further.

The density is considerably higher in Poyang Lake than any other section of Yangtze River system.

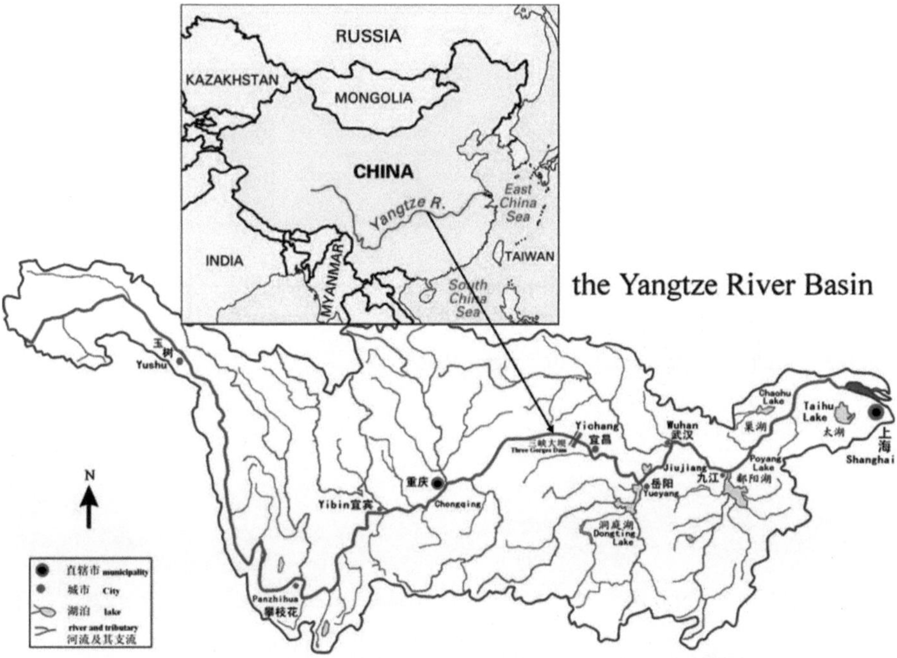

Fig. D.1 Distribution of YFP in lower and middle reaches of Yangtze River

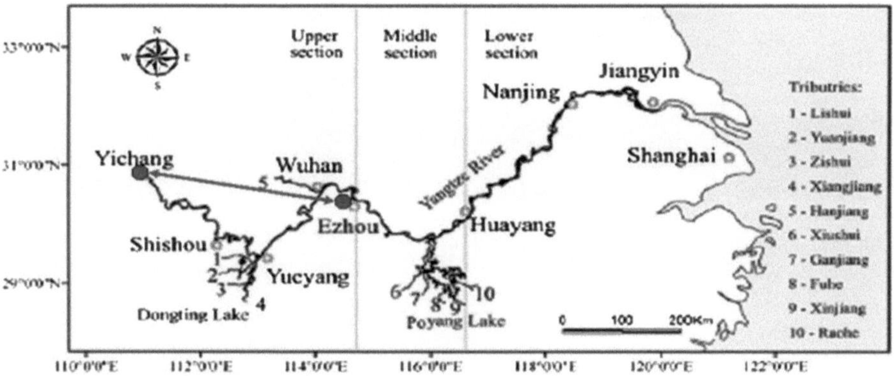

Fig. D.2 Population change of YFPs

D.1.4 Threats to YFPs in Poyang Lake

Sand Mining

- Makes noise ⇒ disturbs the sonar system of porpoises and scares them
- Causes turbidity ⇒ hinders porpoises' navigation

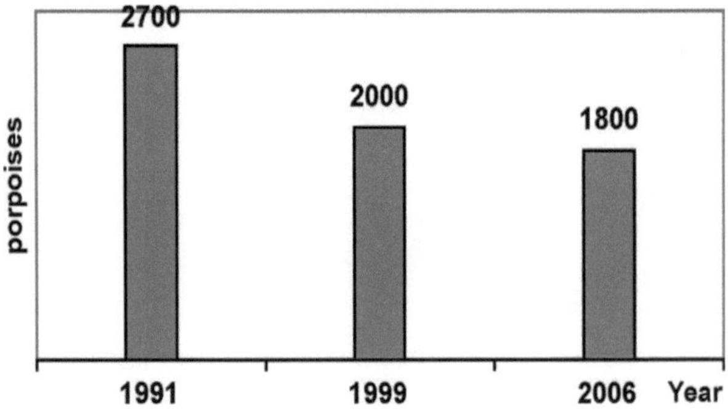

Fig. D.3 Middle and lower reaches of Yangtze River

Fig. D.4 YFPs hurt by fishing gear (Source: Barret 2009)

- Discharges waste ⇒ porpoises need good water quality
- Causes pollution by oil leakage ⇒ porpoises need good water quality
- Occupies habitat ⇒ porpoises need sufficient space

Illegal Fishing and Overfishing

- Causes incidental entanglement in fishing gear (Fig. D.4) Decreases important food sources (fish) for porpoises (Fig. D.5)

Fig. D.5 Overfishing caricature (Source: Society Environment Economy Knowledge (SEEK) 2011)

Industrial and Agricultural Pollution

- Heavy metal pollution ⇒ porpoises need good water quality
- Pesticide ⇒ porpoises need good water quality

Shipping Traffic

- Makes noise ⇒ disturbs the sonar system of porpoises and scares them
- Produces pollution ⇒ porpoises need good water quality
- Hurts porpoises by accident

Water Projects Such as Bridges, Dams, and Dikes

- Fragment porpoises' habitat (Fig. D.6)
- Cuts off the migration route of porpoises
- Enhanced traffic noise

D.1.5 Why Protect YFPs?

- They are unique in the world.
- They are top predators in the food web and play a key role in the balance of community structure (see the right Figure).

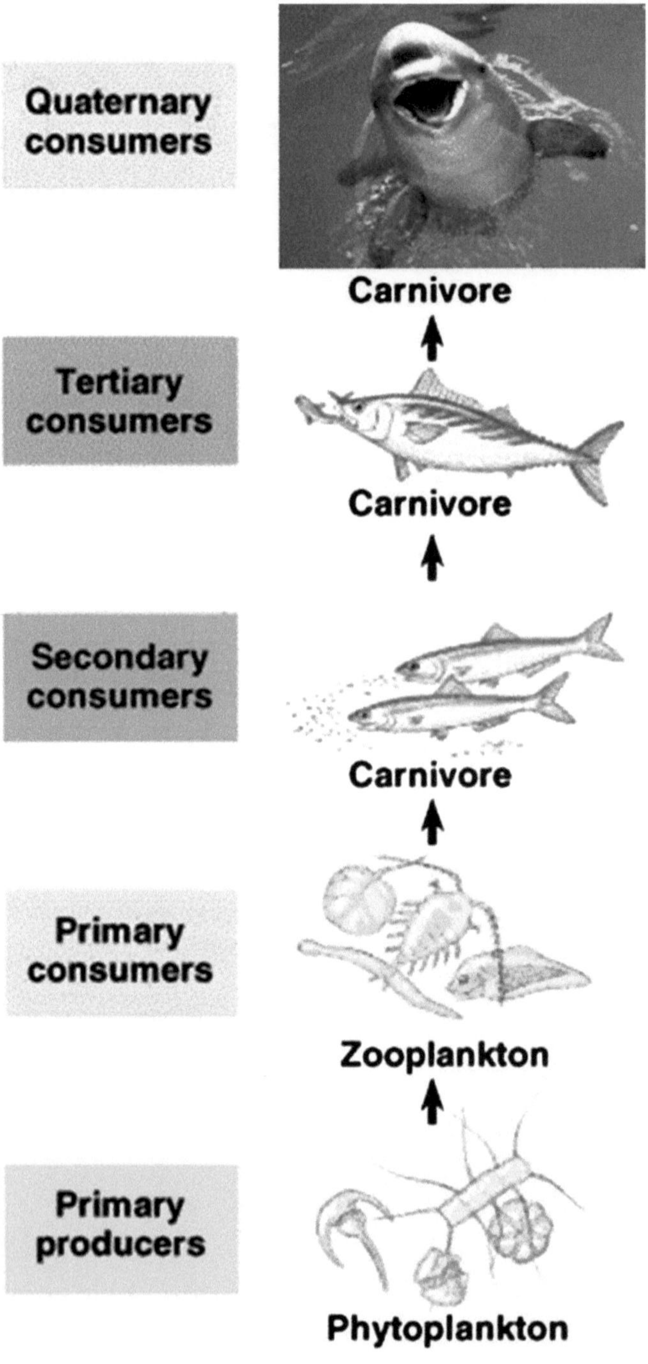
Quaternary consumers
Carnivore
Tertiary consumers
Carnivore
Secondary consumers
Carnivore
Primary consumers
Zooplankton
Primary producers
Phytoplankton

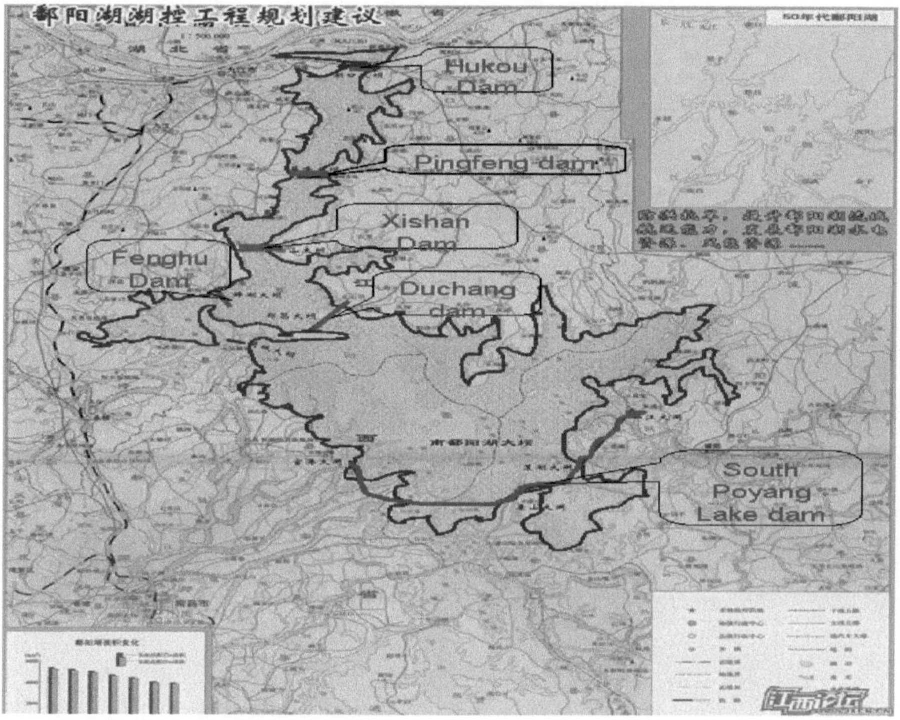

Fig. D.6 Planned dams in Poyang Lake

- They can tell people whether the river environment is suitable for human beings.
- They play an important role in maintaining biodiversity.
- They are the Second Order of National Key Protected Wildlife Species.
- They have been listed as *Endangered Species* on the IUCN Red List since 1996.
- According to "China Wildlife Protection Law," the person who hurts or kills YFPs will receive criminal punishment.
- They are called "giant pandas in Yangtze River" but live in worse situation than pandas.
- They are as important as river dolphins (Baiji), which died out in 2005.
- They are at a very high risk of extinction.

D.1.6 Situation of YFPs in Poyang Lake

~450 porpoises live in Poyang Lake, accounting for 25 % of the total population in China and in the world (Fig. D.7).

Poyang lake is a suitable habitat for finless porpoises because:

- Not much porpoise population change
- Higher porpoise density

Fig. D.7 Poyang Lake

- Many female pregnant porpoises
- There is enough space for a larger population

D.1.7 Possible Future Scenarios of YFPs in Poyang Lake

- If no effective protection measures are taken, the ecology of Poyang Lake will degrade seriously and, therefore, YFPs will become extinct in 25–100 years. It means that we cannot see them any more in the world.
- If some effective measure are taken, such as to build up a special nature reserve for YFPs, the threats can be eliminated and the habitat conditions can be improved greatly. Consequently, the number of YFPs can increase gradually.

Proposed Conservation Project for YFPs in Poyang Lake

Background: Poyang Lake Project is a large-sized hydroelectric project that aims to prevent flooding and droughts, control diseases named Schistosomiasis and produce hydroelectricity (Fig. D.8). At the same time, because of building dams, the habitat of YFPs will be fragmented. Therefore, it is very urgent to take measures on the porpoise protection.

Project planning: At present, the *World Wildlife Fund (WWF)* and *the Wuhan Baiji Conservation Foundation in China* are planning to establish a nature reserve in Poyang Lake for YFPs because biologists think creating a nature reserve is the most effective way for porpoise conservation. In the nature reserve sand mining would be banned and fishing would be regulated. A feasible size and location has been identified for the nature reserve (Fig. D.1). The nature reserve would be 300 km^2, about one third of Hongkong.

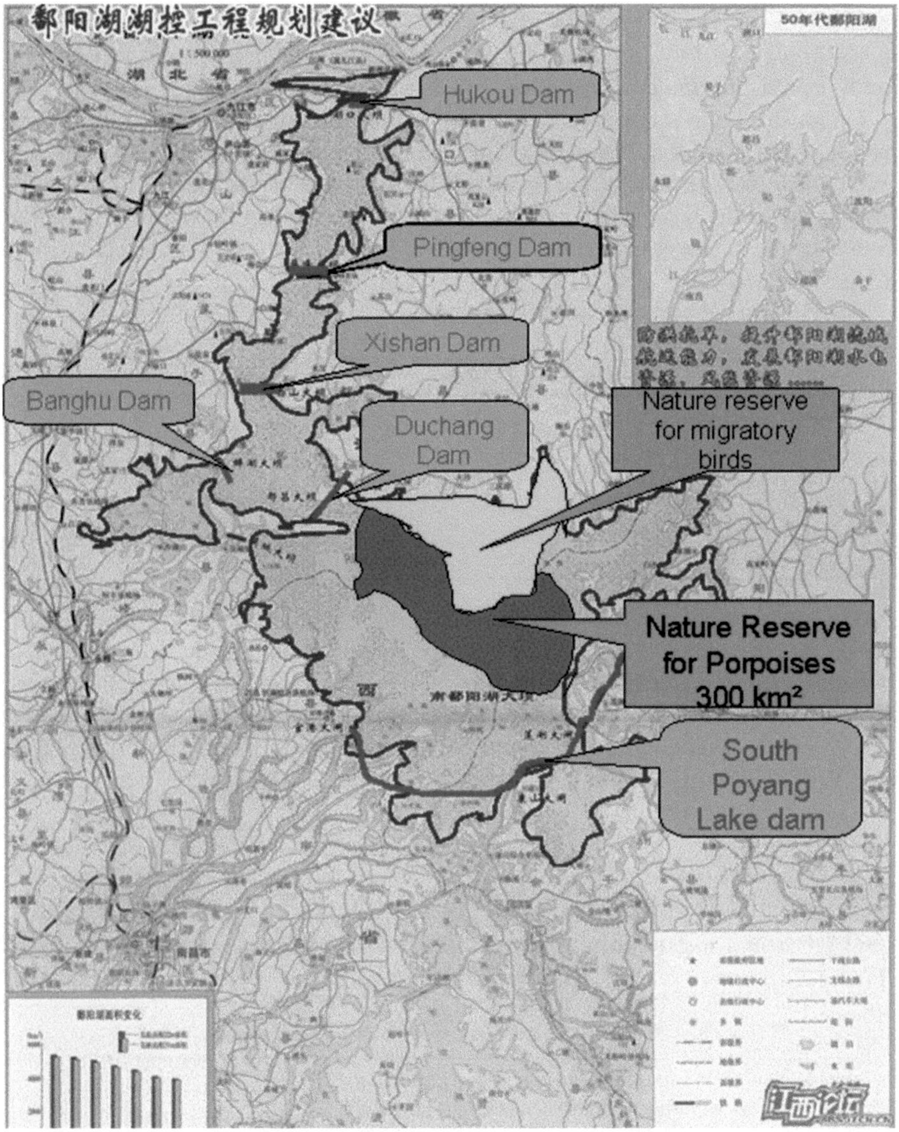

Fig. D.8 Poyang Lake project and proposed nature reserve

In the nature reserve, there would be no risk of entanglement with fishing gear or crash with sand mining machines so that porpoises can navigate freely and safely. That is, the habitat conditions for YFPs would come to a good state so that the porpoises can reproduce smoothly. Since the fragmented parts of Poyang Lake would not be suitable for porpoises, the plan is to transfer the porpoises living in the north of Duchang Dam to the nature reserve.

Project Aim: It is estimated that the porpoise population would increase from 400 to 600 in the next 10 years in the nature reserve. This would mean that YFPs would not be at risk of extinction any more.

Establishing the nature reserve would cost a lot of money (e.g., training and paying locals to work as "rangers"). Since nature conservation organizations are not able to bear the full costs of the project, *WWF* has created a special "***YFP fund***" to support the nature reserve. In order to find out whether enough money can be raised to finance the nature reserve, we would like to know if you would like to contribute to the "***YFP fund***."

Q.1 Do you support the conservation project?

Yes □	No □	Don't know □
Please explain your answer:		

Since *nature conservation organizations are* not able to bear the full costs of the project, a special "*YFP fund*" has been created to support the nature reserve. In order to find out whether enough money can be raised to finance the nature reserve, we would like to know if you would like to contribute to the "*YFP fund.*"

Q.2 Would you be willing to donate some money (even a very small amount) to the "YFP fund"?

Yes	□	If yes, please continue with Q3
No	□	If no, please continue with Q5

Q.3 I would like to contribute to the conservation project on YFPs because (*Please cross your answer among the following options. You can cross several options.*)

I have seen YFPs and I think they are worth protecting	**(1)**
YFPs play an important part in maintaining biodiversity	**(2)**
I would go to Poyang Lake for the sightseeing of YFPs if possible in future	**(3)**
I hope my future generations can enjoy the existence of YFPs	**(4)**
I hope some other people can enjoy the existence of YFPs	**(5)**
I want YFPs to continue live in the world	**(6)**

Q.4 Based on your above reason, what is the most that you would be willing to donate towards "***YFP fund***"?

Before you answer, please remember that

- It is a one-off payment.
- How much can you afford.

- How important the porpoise conservation project is for you.
- The money raised would only go toward protection of YFPs in Poyang Lake.
- If the raised money is not enough, the nature reserve cannot be established.

The most I would be willing to donate for the protection of YFPs in Poyang Lake is:

_____________________RMB

When you feel it difficult to decide, you can refer to the following amounts (please circle one value): (Note: 1 Euro ≈ 9 RMB)

1 RMB □	5 RMB □	10 RMB □	15 RMB □	20 RMB □	25 RMB □	30 RMB □
35 RMB □	40 RMB □	45 RMB □	50 RMB □	55 RMB □	60 RMB □	65 RMB □
70 RMB □	75 RMB □	80 RMB □	85 RMB □	90 RMB □	100 RMB □	

Please explain how you decide on this figure?

Q.5 I would not like to donate any money to the "YFP fund" because *(Please cross your answer among the following options. You can make several options)*

I have no money	**(1)**
It is China government's business	**(2)**
Money should be collected from factories around Poyang Lake	**(3)**
Money should be collected from big companies	**(4)**
NGOs such as WWF, IUCN, should be responsible for their conservation	**(5)**
It is none of my business	**(6)**
I do not care about porpoises	**(7)**
I am afraid that the money will not be used for the conservation project	**(8)**
Other reason:____________________________________	**(9)**

References

Barret, L. (2009). Finless porpoises, Available from http://www.edgeofexistence.org/edgeblog/wp-content/uploads/2009/01/fi nless-porpoise-1.jpg

Society Environment Economy Knowledge (SEEK). (2011). Crisis: Overfishing. Available from http://www.myseek.org/wp-content/uploads/2011/11/problem-of-overfishing.jpg

Index

Y. Dong, *Contingent Valuation of Yangtze Finless Porpoises in Poyang Lake, China*,
DOI 10.1007/978-94-007-2765-6,

MIX
Papier aus verantwortungsvollen Quellen
Paper from responsible sources
FSC® C105338

If you have any concerns about our products,
you can contact us on
ProductSafety@springernature.com

In case Publisher is established outside the EU,
the EU authorized representative is:
Springer Nature Customer Service Center GmbH
Europaplatz 3, 69115 Heidelberg, Germany

Printed by Libri Plureos GmbH
in Hamburg, Germany